# The Strange World of Quantum Mechanics

This is an exceptionally accessible, accurate, and non-technical introduction to quantum mechanics.

After briefly summarizing the differences between classical and quantum behavior, this engaging account considers the Stern–Gerlach experiment and its implications, treats the concepts of probability, and then discusses the Einstein–Podolsky–Rosen Paradox and Bell's theorem. Quantal interference and the concept of amplitudes are introduced and the link revealed between probabilities and the interference of amplitudes. Quantal amplitude is employed to describe interference effects. Final chapters explore exciting new developments in quantum computation and cryptography, discover the unexpected behavior of a quantal bouncing ball, and tackle the challenge of describing a particle with no position. Thought-provoking problems and suggestions for further reading are included.

Suitable for use as a course text, *The Strange World of Quantum Mechanics* enables students to develop a genuine understanding of the domain of the very small. It will also appeal to general readers seeking intellectual adventure.

DAN STYER is Professor of Physics at Oberlin College. A graduate of Swarthmore College and Cornell University, he has published technical research papers in *Physical Review, Journal of Statistical Physics* and the *Proceedings of the Royal Society*. Styer is an associate editor of the *American Journal of Physics*, and his quantum mechanics software won the 1994 Computers in Physics Educational Software Contest. A man of lively intellect, Styer's goal in life is to keep learning new things, and to that end he invests energy into presenting science to a general audience. 'I learn a lot through research and by teaching technical courses to physics majors,' says Styer, 'but I learn even more by distilling the essence of physics ideas into a rigorously honest yet non-technical presentation for a general audience. To reach this group, I cannot hide my ignorance behind a screen of mathematical formulas or technical jargon.' Professor Styer enjoys running, backpacking, and rearing his two children as well as doing science.

# The Strange World of Quantum Mechanics

Daniel F. Styer
*Oberlin College, Ohio*

CAMBRIDGE
UNIVERSITY PRESS

CAMBRIDGE UNIVERSITY PRESS
Cambridge, New York, Melbourne, Madrid, Cape Town, Singapore, São Paulo

Cambridge University Press
The Edinburgh Building, Cambridge CB2 2RU, UK

Published in the United States of America by Cambridge University Press, New York

www.cambridge.org
Information on this title: www.cambridge.org/9780521661041

First published 2000
Fourth printing 2004

*A catalogue record for this publication is available from the British Library*

*Library of Congress Cataloguing in Publication data*
Styer, Daniel F.
The strange world of quantum mechanics / Daniel F. Styer.
p.   cm.
Includes index.
ISBN 0 521 66104 8. – ISBN 0 521 66780 1 (pbk.)
1. Quantum theory.  I. Title
QC174.12.S879   1999
530.12–dc21       99–13559   CIP

ISBN-13  978-0-521-66104-1 hardback
ISBN-10  0-521-66104-8 hardback

ISBN-13  978-0-521-66780-7 paperback
ISBN-10  0-521-66780-1 paperback

Transferred to digital printing 2006

Dedicated to two extraordinary teachers of quantum mechanics:

John R. Boccio and N. David Mermin

*There are more things in heaven and earth, Horatio,*
*Than are dreamt of in your philosophy.*

Hamlet I.v.166

# Contents

# Preface

This book presents the two central concepts of quantum mechanics in such a way that non-technical readers will learn how to work simple yet meaningful problems, as well as grasp the conceptual bizarreness of the quantal world. Those two central concepts are: (1) The outcome of an experiment cannot, in general, be predicted exactly; only the probabilities of the various outcomes can be found. (2) These probabilities arise through the interference of amplitudes.

The book is based on a short course (only fourteen lectures) that I have presented to general-audience students at Oberlin College since 1989, and thus it is suitable for use as a course textbook. But it is also suitable for individual readers looking for intellectual adventure. The technical background needed to understand the book is limited to high school algebra and geometry. More important prerequisites are an open mind, a willingness to question your ingrained notions, and a spirit of exploration. Like any adventure, reading this book is not easy. But you will find it rewarding as well as challenging, and at the end you will possess a genuine understanding of the subject rather than a superficial gloss.

How can one present a technical subject like quantum mechanics to a non-technical audience? There are several possibilities. One is to emphasize the history of the subject and anecdotes about the founders of the field. Another is to describe the cultural climate, social pressures, and typical working conditions of a quantum physicist today. A third is to describe useful inventions, such as the laser and the transistor, that work through the action of quantum mechanics. A fourth is to outline in general terms the mathematical machinery used by physicists in solving quantum mechanical problems.

I find all four of these approaches unsatisfactory because they emphasize quantum *physicists* rather than quantum *physics*. This book uses instead a fifth approach, which emphasizes how nature behaves rather than how humans behave. Humans have certainly been very clever in discovering and using quantum mechanics, and I am proud of our species for its activities in this regard. But in this book (except for the appendices) the focus rests squarely on nature and not on how we study nature.

In order to solve problems in quantum mechanics, the professional physicist has erected a gigantic and undoubtedly elegant mathematical edifice. This edifice is necessary for finding the answers to specific problems (which is, after all, what physicists are paid to do), but it often conceals rather than reveals the underlying physical principles of quantum mechanics. Physicists, in fact, are often clumsy in their use and understanding of quantum mechanics's central concepts; they are protected from them by a screen of mathematics. (The very name "*quantum mechanics*" memorializes an aspect of atomic physics that is not central to quantum mechanics and that appears in the classical world as well.) This book aims to strip away the machinery of the edifice and bare the raw ideas in their naked form.

An analogy helps to explain this aim. The professional automobile mechanic must be familiar with crankshafts and camshafts, pistons and plugs, transmissions and timing. His familiarity enables him to repair cars and earn his salary. Yet these practical and interesting devices are irrelevant to the central concept of how a car works — which is simply that hot air expands, whence heat from burning gasoline can be converted into motion. Many excellent mechanics are in fact unfamiliar with this central concept. A book on the fundamental workings of automobiles would discuss heat and motion, but would not tell you how to give your car a tune-up. You should expect analogous discussions here: no more and no less.

Above I have described the direct goals of this book. Two other goals are indirect yet just as important. First, I aim to describe scientific thought — its character, its strengths, its limitations — and to inspire an appreciation for the elegance, economy, and beauty of scientific explanations. Second, I hope to demonstrate the importance and power of reason as a tool for solving problems and probing the unknown. The popular press is fond of misstatements like "the belief in an objective reality, accessible to reason, ... suffered a death blow with the advent of modern physics".* The truth is that quantum mechanics is unfamiliar, non-common-sensical, and weird, but it is perfectly logical and rational. Indeed, in the bizarre world of quantum mechanics, it is logic, and not common sense, that is the only sure guiding light. In today's cultural atmosphere — where in-your-face power play has largely displaced rational debate in the arena of public discourse — this point cannot be overemphasized.

---

* Sources for direct quotations are gathered in appendix C on page 138.

This book describes quantum mechanics as most physicists understand it today. All scientific knowledge is tentative and the pillars of quantum mechanics are no exception. In addition, the experiments and principles described here are all subject to interpretation. I present the standard interpretation, which is not the only one. (I give only fleeting mention to alternative interpretations and formulations not because they are incorrect or unimportant, but because one must have a firm grasp on the standard interpretation before moving on to the alternatives.)

> *Technical aside:* Sometimes it is useful to make a point that is rather technical and that is not essential for developing the book's argument. Such technical asides are labelled and indented, like this sample.

Producing a completely honest yet non-technical account of quantum mechanics is an audacious enterprise, and while developing this treatment I have reached out for help from many people. I need to thank first the 985 Oberlin College students who have, since 1989, taken the course which led to this book. Their questions, objections, doubts, excitement, enthusiasm, and triumphs have inspired many changes — improvements, I hope — in the content and presentation given here, as well as in my own understanding of quantum mechanics. In the spring of 1996 I served as associate instructor for the computer conference course "Demystifying Quantum Mechanics", developed and taught by Edwin F. Taylor. Working with Professor Taylor and the fifteen intrepid students in that class (mostly high school teachers scattered across the nation) was a pleasure that further refined my understanding and this book's presentation.

I received helpful direct comments on this treatment from many of the students mentioned above, and from Gary E. Bowman, Amy Bug, Peter Collings, Rufus Neal, Joe Palmieri, Robert Romer, Dan Sulke, Edwin F. Taylor, and four anonymous reviewers. This is not to say that all of these readers approve of everything I say here — indeed, I know that some of them disagree with me on important points — but I appreciate the contribution that each one of them has made to this work. The illustrations were skillfully drawn by Byron Fouts.

The development of the course which led to this book was supported by a grant from the Sloan Foundation. This acknowledgement may sound like the bland gratitude of someone merely content to receive Sloan's money, but it is not. The encouragement of the foundation, and in particular of program officer Samuel Goldberg, led me to delve deeply into quantum mechanics as a set of physical ideas rather than as an elaborate and somewhat mystical algorithm for solving problems in atomic physics. I have learned much in preparing this account, and

I thank the Sloan Foundation for suggesting that someone other than myself would be interested in what I learned.

I invite you to join the community that has developed this approach and this book. If you have access to the Internet you can send me computer mail at address

Dan.Styer@oberlin.edu

and you will find a World Wide Web page devoted to this book at

http://www.oberlin.edu/physics/dstyer/StrangeQM/

Comments on paper are just as welcome, and should be addressed to

Dan Styer
Physics Department
Oberlin College
Oberlin, Ohio 44074–1088 USA

I offer you my welcome and my best wishes. Enjoy!

# 1
# Introduction

## 1.1 Capsule history of quantum mechanics

Starting in the seventeenth century, and continuing to the present day, physicists developed a body of ideas that describe much about the world around us: the motion of a cannonball, the orbit of a planet, the working of an engine, the crack of a baseball bat. This body of ideas is called *classical mechanics*.

In 1905, Albert Einstein realized that these ideas didn't apply to objects moving at high speeds (that is, at speeds near the speed of light) and he developed an alternative body of ideas called *relativistic mechanics*. Classical mechanics is wrong in principle, but it is a good approximation to relativistic mechanics when applied to objects moving at low speeds.

At about the same time, several experiments led physicists to realize that the classical ideas also didn't apply to very small objects, such as atoms. Over the period 1900–1927 a number of physicists (Planck, Bohr, Einstein, Heisenberg, de Broglie, Schrödinger, and others) developed an alternative *quantum mechanics*. Classical mechanics is wrong in principle, but it is a good approximation to quantum mechanics when applied to large objects.

## 1.2 What is the nature of quantum mechanics?

I'm not going to spend any time on the history of quantum mechanics, which is convoluted and fascinating. Instead, I will focus on the ideas developed at the end. What sort of ideas required twenty-eight years of development from this stellar group of scientists?

Einstein's theory of relativity is often (and correctly) described as strange and counterintuitive. Yet, according to a widely used graduate level text,

1

[the theory of relativity] is a modification of the structure of mechanics which must not be confused with the far more violent recasting required by quantum theory.

Murray Gell-Mann, probably the most prominent living practitioner of the field, said of quantum mechanics that

Nobody feels perfectly comfortable with it.

And the inimitable Richard Feynman, who developed many of the ideas that will be expounded in this book, remarked that

I can safely say that nobody understands quantum mechanics.

One strange aspect of quantum mechanics concerns predictability. Classical mechanics is *deterministic* — that is, if you know exactly the situation as it is now, then you can predict exactly what it will be at any moment in the future. Chance plays no role in classical mechanics. Of course, it might happen that the prediction is very difficult to perform, or it might happen that it is very difficult to find exactly the current situation, so such a prediction might not be a practical possibility. (This is the case when you flip a coin.) But in principle any such barriers can be surmounted by sufficient work and care. Relativistic mechanics is also deterministic. In contrast, quantum mechanics is *probabilistic* — that is, even in the presence of exact knowledge of the current situation, it is impossible to predict its future exactly, regardless of how much work and care one invests in such a prediction.

Even stranger, however, is quantum mechanical *interference*. I cannot describe this phenomenon in a single paragraph — that is a major job of this entire book — but I can give an example. Suppose a box is divided in half by a barrier with a hole drilled through it, and suppose an atom moves from point P in one half of the box to point Q in the other half. Now suppose a second hole is drilled through the barrier and then the experiment is repeated. The second hole increases the number of possible ways to move from P to Q, so it is natural to guess that its presence will increase the probability of making this move. But in fact — and in accord with the predictions of quantum mechanics — a second hole drilled at certain locations will *decrease* that probability.

The fact that quantum mechanics is strange does not mean that quantum mechanics is unsuccessful. On the contrary, quantum mechanics is the most

successful theory that humanity has ever developed; the brightest jewel in our intellectual crown. Quantum mechanics underlies our understanding of atoms, molecules, solids, and nuclei. It is vital for explaining aspects of stellar evolution, chemical reactions, and the interaction of light with matter. It underlies the operation of lasers, transistors, magnets, and superconductors. I could cite reams of evidence backing up these assertions, but I will content myself by describing a single measurement. One electron will be stripped away from a helium atom that is exposed to ultraviolet light below a certain wavelength. This threshold wavelength can be determined experimentally to very high accuracy: it is $50.4259299 \pm 0.0000004$ nanometers. The threshold wavelength can also be calculated from quantum mechanics: this prediction is $50.4259310 \pm 0.0000020$ nanometers. The agreement between observation and quantum mechanics is extraordinary. If you were to predict the distance from New York to Los Angeles with this accuracy, your prediction would be correct to within the width of your hand. In contrast, classical mechanics predicts that *any* wavelength of light will strip away an electron, that is, that there will be no threshold at all.

## 1.3    How small is small?

I said above that the results predicted by quantum mechanics differed from the results predicted by classical mechanics only when these ideas were applied to "very small objects, such as atoms". How small is an atom? Cells are small: a typical adult contains about 60 trillion cells. But atoms are far smaller: a typical cell contains about 120 trillion atoms. An atom is twice as small, relative to a cell, as a cell is small, relative to a person. In this book, when I say "small" I mean "*very* small". You've never handled objects this small; I've never handled objects this small; none of your friends has ever handled objects this small. They are completely outside the domain of our common experience. As you read this book, you will find that quantum mechanics is contrary to common sense. There is nothing wrong with this. Common sense applies to commonly encountered situations, and we do not commonly encounter the atomic world.

## 1.4    The role of mathematics in quantum mechanics

One frequently hears statements to the effect that the ideas of quantum mechanics are highly mathematical and can only be understood through the use of complex mathematics (partial differential equations, Fourier transforms, eigenfunction expansions, etc.).

> One can popularize the quantum theory [only] at the price of gross oversimplification and distortion, ending up with an uneasy compromise between what the facts dictate and what it is possible to convey in ordinary language.

It is certainly true that the professional physicist needs a vast mathematical apparatus in order to solve efficiently the problems of quantum mechanics. (For example, the calculation of the helium stripping threshold wavelength described above was a mathematical *tour de force*.) But this is not, I believe,* because quantum mechanics itself is fundamentally difficult or mathematical. I believe instead that the root rules of quantum mechanics are in fact quite simple. (They are unfamiliar and unexpected, but nevertheless simple.) When these rules are applied to particular situations, they are used over and over again and therefore the *applications* are complicated. An analogy helps explain this distinction. The rules of chess are very simple: they can be written on a single page of paper. But when these rules are applied to particular situations they are used over and over again and result in a complicated game: the applications of the chess rules fill a library.

Indeed, can any fundamental theory be highly mathematical? Electrons know how to obey quantum mechanics, and electrons can neither add nor subtract, much less solve partial differential equations! If something as simple-minded as an electron can understand quantum mechanics, then certainly something as wonderfully complex as the reader of this book can understand it too.

---

* Not everyone agrees with me.

# 2
# Classical Magnetic Needles

How shall we approach the principles of quantum mechanics? One way is simply to write them down. In fact I have already done that (in the first paragraph of the Preface), but to do so I had to use words and concepts that you don't yet understand. To develop the necessary understanding I will use a particular physical system as a vehicle to propel our exploration of quantum mechanics. Which system? An obvious choice is the motion of a tossed ball. Unfortunately this system, while simple and familiar in classical mechanics, is a complicated one in quantum mechanics. We will eventually get to the quantum mechanics of a tossed ball (in chapter 14, "Quantum mechanics of a bouncing ball", page 103), but as the vehicle for developing quantum mechanics I will instead use a system that is simple in quantum mechanics but that is, unfortunately, less familiar in daily life. That system is the magnetic needle in a magnetic field. This chapter describes the classical motion of a magnetic needle so that we will be able to see how its classical and quantal behaviors differ.

## 2.1  Magnetic needle in a magnetic field

A magnetic needle — like the one found in any woodsman's compass — has a "north pole" and a "south pole". I will symbolize the magnetic needle by an arrow pointing from its south pole to its north pole. When a magnetic needle is placed in a magnetic field — such as the magnetic field of the earth, or that produced by a horseshoe magnet — then the magnetic field acts to push the north pole in the direction of the field, and to push the south pole in the direction opposite the field. (It is not important for you to understand in detail how this effect works or even what the phrases "north pole" and "magnetic field" mean. Remember that this chapter merely builds a classical scaffolding that will be discarded once the correct quantal structure is built.) These two pushes together

twist the needle towards an orientation in which the associated arrow
points in the same direction as the field. If the needle starts out pointing
parallel to the magnetic field, then it keeps on pointing in that direction.
If the needle starts out not pointing parallel to the magnetic field, then
it oscillates back and forth about this preferred direction. (If friction
is present, then these oscillations will eventually die out and the needle
will point precisely parallel to the field. If there is no friction then the
oscillations will continue forever. In atomic systems there is no friction.)

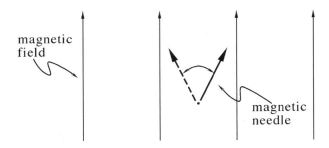

If the magnetic field has the same strength at all points in the vicinity
of the needle, that is, if the field is uniform, then the upward force acting
on the north pole of the needle is exactly cancelled by the downward
force acting on the south pole and there is no net force on the needle. So
in a uniform field there is an impetus for the needle to oscillate, but no
impetus for it to move up or down, or left or right.

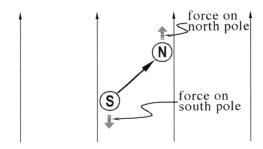

## 2.2   Magnetic effects on electric current

A loop of wire carrying electric current behaves in many ways like a
compass needle. The associated magnetic arrow* points perpendicular to
the current loop, so if the current loop is placed in a magnetic field, the

---

* This associated arrow is purely abstract — there's nothing actually located there.

arrow "wants" to point parallel to the field. But the current loop's arrow

isn't *exactly* like a compass needle's arrow, because the current loop arrow *precesses* rather than *oscillates* in a magnetic field. "Precession" means that the tip of the symbolic arrow moves around a circle while its base is fixed. Thus a precessing arrow traces out the figure of a cone. You

can make your index finger precess by holding it up in the air and then twisting its tip around in a circle while keeping your hand fixed.

I wish I could describe for you an experiment that you could do to prove this fact to yourself. Unfortunately, this cannot be done with the equipment available in the typical home. It is, however, quite easy to do a parallel home experiment with an analogous system. A top rotating in a gravitational field happens to behave very much like a current loop in a magnetic field. (The rotating body of the top is analogous to the moving electric charge — the current — in the loop. The axle of the top is analogous to the magnetic arrow.) I urge you to spin a top, put it on the floor, tip the rotation axis away from the vertical, and then watch the top precess.

## 2.3 Magnetic needle in a non-uniform magnetic field

We have seen that a magnetic needle in a uniform magnetic field feels zero net magnetic force, because the upward force on the north pole is cancelled by the downward force on the south pole. But if a magnetic needle is placed in a *non-uniform* magnetic field, then there *can* be a net force on the needle.

The figure below shows a magnetic field which is stronger at the top of the figure than at the bottom of the figure. For the horizontal needle, both the north and south poles are at the same height and experience the same magnetic field strength, so the two poles experience equal but opposite forces and the net force vanishes. But for the vertical needle, the north pole experiences a stronger magnetic field than does the south pole, so there is a larger upward force on the north pole and a smaller downward force on the south pole. As a result the two forces don't completely cancel — there remains a net upward force. The tilted needle is intermediate between these two situations. It experiences a net upward force, but that force is not as strong as the force on the vertical needle.

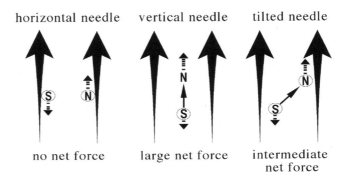

horizontal needle    vertical needle    tilted needle

no net force    large net force    intermediate
net force

You can see that the net force depends upon the angle between the arrow and the field. In fact, the force is proportional to a quantity bearing the awkward name of "the projection of the magnetic arrow onto the direction of the magnetic field". This quantity is defined through a four-stage process: (1) Draw a line to show the direction of the magnetic field (in the illustration below, it tilts to the left). (2) Draw in the magnetic arrow with its base on the field line. (3) Draw a line perpendicular to the field line through the base of the arrow, and another through the tip of the arrow (these are shown dashed). (4) The distance between these two lines is the desired "projection".

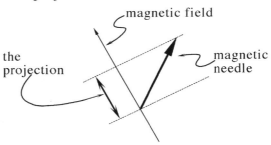

magnetic field

the
projection

magnetic
needle

Examples of projections:

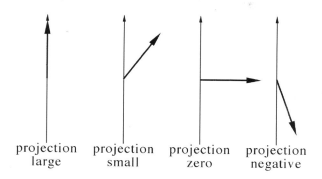

If an electric current loop is placed in a non-uniform magnetic field, its arrow will precess and at the same time the loop will move. During this precession the projection remains constant,[†] and hence the force remains constant. For example, suppose the field is stronger at the top than at the bottom (as in the figure on page 8) and suppose a stationary current loop with a small positive projection is placed into the field. Then the current loop will move upward, and as it moves it precesses in such a way that the impetus to move upward stays constant. If the initial projection is negative, then the current loop moves downward.

## 2.4   Explanation *vs.* description

Have I *explained* the motion of magnetic needles in magnetic fields? Have I *explained* the nature of a magnetic field? Not at all! I have simply *described* these phenomena. Sometimes a description in science can be explained through an appeal to more fundamental principles. For example, I have spoken about the north and south magnetic poles of a compass needle. The poles of a compass needle can in fact be explained in terms of the motion of electrons within the needle's atoms. But in other cases the description is simply the most fundamental thing there is and cannot be "explained" by something else. What is a magnetic field? I have described it, in essence, as "that which makes a compass needle want to oscillate". There are more elaborate and more mathematical descriptions of magnetic field, but none are more fundamental. Science has no explanation for magnetic field, only a description of it.

---

[†] Spin your top again and notice that as the top precesses, the tip of the axle remains always the same distance from the floor. The vertical distance from the floor to the tip of the axle is the projection of the axle onto a vertical line. (If you wait long enough that friction slows the rotation of the top, then this projection — the height of the axle tip — will decrease. But if friction can be ignored, then the projection does not change.)

What does "explanation" mean, anyway? Suppose you ask me "Why did it rain yesterday?" I might reply "Because a cold front moved in." Then you could ask "But why did a cold front move in?" I might say "Because the jet stream pushed it." You: "But why did the jet stream push it?" Me: "Because the sun warmed Saskatchewan and so deflected the jet stream."‡ You: "But why does sunlight warm objects?" And at this level I really can't answer your question. I know *that* sunlight carries energy (so do you), and science can describe this energy transport with exquisite accuracy. But science cannot *explain* this energy transport or tell *why* it happens.

This story illustrates that "explanation" means "explanation in terms of something more fundamental". At some point any chain of questioning descends to the most fundamental ideas, and there it must stop. Currently, the most fundamental ideas in physics are called "quantum electrodynamics" and "quantum chromodynamics", two theories which fall squarely within the framework of quantum mechanics that I will describe in this book. Probably there will someday be even more fundamental ideas, so that "why" questions concerning quantum electrodynamics could be answered in terms of these new ideas. However, "why" questions concerning these more fundamental ideas will then be unanswerable! Ultimately, at the bottom of any descending chain of questions, science can only give descriptions (facts) and not explanations (reasons for those facts).

## 2.5   Problems

*Above all things we must beware of what I will call "inert ideas" — that is to say, ideas that are merely received into the mind without being utilized, or tested, or thrown into fresh combinations.*

— Alfred North Whitehead

Reading books, listening to lectures, watching movies, running computer simulations, performing experiments, participating in discussions ... all these are fine tools for learning quantum mechanics. But you will not *really* become familiar with the subject until you get it under your skin by working problems. The problems in this book do not simply test your comprehension of the material you read in the text. They are instead an important component of the learning process, designed to extend and solidify the concepts presented. Solving problems is a more active, and

---

‡ Anyone who has raised a child is all too familiar with such chains of questions.

hence more effective, learning technique than reading text or listening to lectures.

Some might contend that problems have no place in a book intended for a general audience, because they are "too hard". In fact the opposite is true: it is easier to learn by working problems than by reading words. If "working problems" seems too dry or too regimented for you, then think of it as "solving puzzles" instead.

If you write up solutions to these and subsequent problems (such as for a course assignment) be sure to explain your reasoning. Don't just write down the final numerical answer — your teacher already knows what it is! Instead (s)he is interested in seeing how you overcome the roadblocks that get in your way as you progress through the problem. Appendix F (page 149) contains skeleton answers for some of the problems. (There are also three complete sample solutions on pages 28, 71, and 109.) By a "skeleton answer", I mean only the "final numerical answer" mentioned above without any of the reasoning that leads to the answer. I do not present the reasoning because the benefits that accrue from active problem solving come only if you supply that reasoning yourself. The appendix will help you learn quantum mechanics if you work through the problem yourself and then use the skeleton answer to check your reasoning. If you instead look up the answer before attempting the problem, the appendix will actually be an impediment to your learning.

Many of the problems ask for short verbal answers. In all such cases, the answer can be written in four or fewer sentences. If you find yourself writing an extended essay, then you either misunderstood the question or don't know the answer. In neither case will your teacher be impressed by the mere bulk of your response.

> *Technical aside:* This book is intended for a general audience, but it is useful also for students of physics and chemistry who can perform calculations far more sophisticated than anything mentioned here, but who are at a loss to explain what it is that they are calculating. For such readers I have included a few problems that require a more technical background: these problems are clearly marked. Such problems are not harder than regular problems, they just require a background knowledge of physics ideas that general readers are unlikely to possess.

2.1 *Variously tilted needles.* Consider the non-uniform magnetic field of the figure on page 8. Describe the net force acting upon a vertical magnetic needle that points downward, and a tilted magnetic needle that points downward and to the left.

2.2 *Projections on a vertical axis.* The figure below shows four magnetic arrows, labeled A, B, C, and D. All four arrows have the same length. Rank them from highest value of the projection to the lowest. (Do not ignore sign. For example, a large negative projection should rank below a small positive projection.) If two of the projections are equal, then say so.

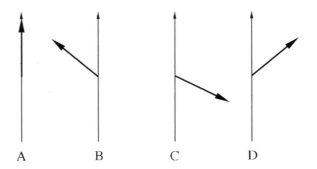

2.3 *Projections in geography.* The radius of the earth is 3960 miles. The Old Mission Point Lighthouse near Traverse City, Michigan, is located at a latitude of 45°. Imagine an arrow extending from the center of the earth to the Old Mission Lighthouse. How long is the projection of this arrow onto the earth's rotation axis? Hint: Recall a geometrical result about a $1:1:\sqrt{2}$ right triangle.

2.4 *Projections on a non-vertical axis.* Up to now we have emphasized projections onto a vertical axis. But our definition applies to *any* axis. In the figure below, find the projection of the short, thick arrow on the long, thin axis. Give your answer in inches, with a + or − sign.

2.5 *Different projections on different axes.* Show that for any arrow, you can pick an axis such that the projection of the arrow on that axis is zero. How many such axes are there?

2.6 *The role of mathematics in quantum mechanics.* One of my students wrote "If you can't read music, then you can't write it, but that doesn't mean you can't understand it." To what extent is this analogy appropriate to the use of mathematics in physics?

# 3

# The Stern–Gerlach Experiment

## 3.1   Measuring magnetic projections

What does the previous chapter have to do with quantum mechanics?
I have said that the predictions of quantum mechanics are significantly
different from those of classical mechanics only when applied to very
small objects. How could we make such a tiny compass needle? In fact we
don't need to make one, because nature itself supplies one. It is natural to
suppose that an atom acts like a tiny magnetic needle because its orbiting
electron mimics a current loop.

In 1922, physicists Otto Stern and Walther Gerlach decided to test this
supposition by measuring the magnetic arrow associated with a silver
atom. It is clear that they could not do this by watching an individual
atom precess in a uniform magnetic field! Instead, they injected a moving
silver atom into a non-uniform field and noticed how the resulting force
pushed the atom around. The "Stern–Gerlach apparatus" sketched on the
next page thus measures the projection of an atom's magnetic arrow on
the vertical axis.

What results would you expect from this experiment? Think about this
for a moment before reading on.

## 3.2   Classical expectations

I don't know about you, but here is what I would expect: Once the atom
enters the non-uniform magnetic field, its magnetic arrow precesses in
such a way that its projection on the vertical axis remains constant. While
this precession is taking place, there is also a force on the atom, and the
magnitude of that force depends upon the value of the projection. If the
atom has a large positive projection, it will experience a large upward
force and move up sharply. If the atom has a small positive projection,

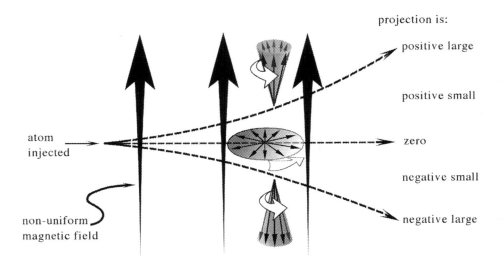

Fig. 3.1.  A sketch of the Stern–Gerlach apparatus, with some justification for my classical expectations.

it will experience a small upward force and move up moderately. If the projection happens to be zero, the atom will experience zero force and move straight through. Similarly for atoms with negative projections.

Thus an atom that happened to enter the field with its arrow pointing straight up ("toward the north pole") would experience a large upward deflection. One that happened to enter with its arrow pointing straight down ("toward the south pole") would experience a large downward deflection. And one that happened to enter with a horizontal arrow ("toward the equator") would experience no deflection. Atoms whose arrows had intermediate tilts would experience intermediate deflections.

Now, there is only one way for an arrow to point toward the north pole, and only one way for it to point toward the south pole, but there are lots of ways for it to point toward the equator.* There are, in fact, a few ways to point toward the 10° north latitude line, more ways to point toward the 20° north latitude line, still more for the 30° line, and so on until a maximum is reached at the equator (the 90° line). I expect atoms to enter the apparatus with their magnetic arrows pointing every-which-way: some straight up, some straight down, most somewhere in between. Thus I expect a very small number of atoms to come out with the maximum upward deflection, a larger number to come out with moderate upward deflections, the largest number to come out with zero deflection, and sim-

---

* For example, directly to the right, directly to the left, directly out of the page, half-way between "to the right" and "out of the page", etc.

ilarly for downward deflections. In short, my classical expectation is that
the number of atoms leaving the apparatus with a given deflection should
depend upon the deflection in the manner sketched here.

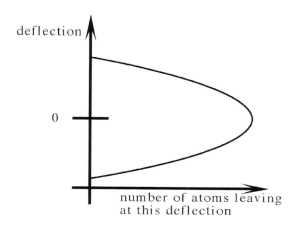

(These expectations are for an ideal experiment. In any real experiment
things go wrong — the magnetic field has small imperfections, the source
of atoms is not perfectly pure, an atom hits a piece of dust while traveling
through the apparatus — so I expect that for a real experiment the curve
obtained will be somewhat broadened and stretched away from the results
shown above.)

### 3.3    Actual results

Imagine how surprised Stern and Gerlach must have been when they
obtained results that were nothing like the expectations described above.
They found that no silver atoms at all went straight through the apparatus.
Nor was there a gradual change in number of exiting atoms with deflection.
In fact, they found that all of the atoms came out at just two different
deflections: one a certain amount up, the other the same amount down.
The observed results for silver atoms are summarized in the graph on
the next page, where the width of the two humps is due entirely to
imperfections in the apparatus.

When atoms other than silver were put through a Stern–Gerlach apparatus, there were sometimes four or five narrow humps, sometimes even
more, but never the broad curve of our classical expectations. Furthermore, when the observed deflections were used to compute the values
of the magnetic projections, then in all cases, from all different kinds of
atoms, the value of that projection turned out to be an integer times a certain quantity called the "Bohr magneton", $m_B = 9.27 \times 10^{-24}$ joule/tesla.

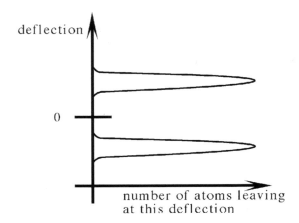

With silver, for example, the two measured values were $+m_B$ and $-m_B$. For nitrogen the four measured values were $+3m_B$, $+m_B$, $-m_B$, and $-3m_B$. For sulfur they were $+4m_B$, $+2m_B$, $0$, $-2m_B$, and $-4m_B$. And so on for other atoms.

*Technical aside:* What if we injected into the apparatus not atoms, but the needles of real live scout compasses? In this case the magnetic projection is huge on an atomic scale — about 0.1 joule/tesla. Presumably the needles will only come out at discrete deflections corresponding to projections of (integer) × $m_B$, but instead of giving rise to four or five narrow humps, there will be about $10^{22}$ of them. There are so many humps, and they are (on the scale of a scout compass) so close together, that the individual humps cannot be distinguished. We find instead a washed out pattern very similar to that of our classical expectations. The principle that when quantum mechanics is applied to big things it must give nearly the same result as classical mechanics is called "the correspondence principle" or "the classical limit of quantum mechanics".

### 3.4   Actual experiments

I have described the Stern–Gerlach experiment in the simplest possible way so as to focus your attention on the fundamental parts of the experiment rather than on the mundane parts necessary for its operation. But you should realize that this experiment (like any other experiment) is a lot more complicated and a lot more difficult to carry out than the conceptual outline given above. Compare the photo of a real Stern–Gerlach apparatus on the next page to the conceptual outline sketched in figure 3.1!

Fig. 3.2. A real Stern–Gerlach apparatus (courtesy of Melvin Daybell).

For example, in our discussion we just said "give us a non-uniform magnetic field" and then we drew it on the page. In the laboratory life is more difficult. Stern and Gerlach had to magnetize two large pieces of iron and carefully shape the pieces so that they would produce the desired magnetic field.

We just said that we needed a source of atoms and a detector of atoms. Stern and Gerlach had to build an electrical oven to eject vaporized silver atoms, and they had to design a suitable detector. (For a detector, they used a glass plate placed to the right of the magnets, and injected enough vaporized atoms that they built up a visible silver deposit on the glass. See figure 3.3.)

We didn't mention at all the possibility that while a silver atom was flying through the magnetic field, it might collide with an oxygen molecule and scatter in some random direction. But Stern and Gerlach had to consider the possibility, so they performed the experiment in a vacuum chamber.

Of course, Stern and Gerlach needed instruments to measure the

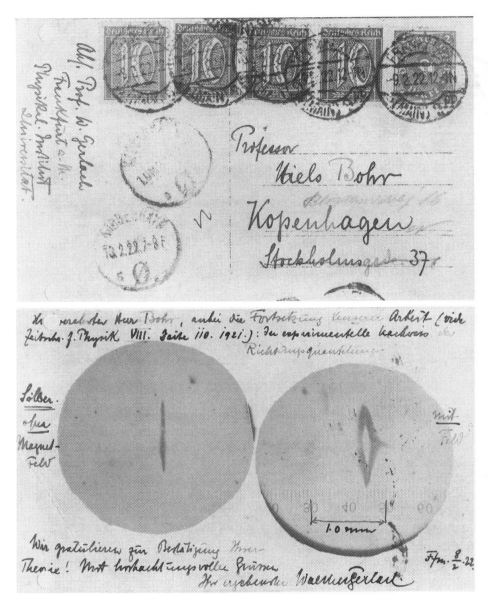

Fig. 3.3. Postcard from Walther Gerlach to Niels Bohr, showing results from one of the earliest, crude Stern–Gerlach experiments. On the left is the beam profile without magnetic field, on the right is the beam profile with a non-uniform magnetic field. Only in the center of the image is the field non-uniformity great enough to pull the two outgoing beams apart. Translation of the message: "My esteemed Herr Bohr, attached is the continuation of our work (vide *Zeitschr. f. Phys.* VIII 110, 1921): the experimental proof of directional quantization. We congratulate you on the confirmation of your theory! With respectful greetings, Your most humble Walther Gerlach." (Courtesy of the Niels Bohr Archive, Copenhagen.)

strength of the magnetic field, the temperature of the oven, and the quality of the vacuum, as well as the number of atoms coming out at a given deflection.

I will mention a number of experiments in this book, and in every case I will present only the simplest conceptual outline. This will keep the concepts clear, but it will ignore a wealth of detail which, while necessary for performing the experiment, serves only to hide the concept. You should be aware that real experiments are always considerably more difficult to perform than thought experiments.

## 3.5   Visualization

Faced with the unexpected results of the Stern–Gerlach experiment, it is natural to seek a reason for these results: to find a picture that tells us what's going on. There is nothing to be ashamed of in this desire. Human beings are visual animals, and we think best in terms of some picture or visualization that we carry in our minds. Nevertheless I urge you to postpone this quest for a visualization. We will first spend considerable time addressing the question: "We know that silver atoms don't behave exactly like miniature compass needles. Just how do they behave?" Once we know the facts about silver atoms, we will try again (in section 15.2, "What does an electron look like?") to produce an accurate visualization. Seeking a visualization at this point, with our incomplete knowledge, will surely produce a mistaken image. Thomas Huxley described the attitude I am advocating by saying:

> Sit down before fact as a little child, be prepared to give up every preconceived notion, follow humbly wherever and to whatever abysses nature leads, or you shall learn nothing.

## 3.6   References

A computer program to simulate the Stern–Gerlach experiment is

> Daniel V. Schroeder, *Spins.*

You may download this free program (it works on Macintosh computers) through the World Wide Web site mentioned on page xiv.

The history of the Stern–Gerlach experiment is traced in

> Immanuel Estermann, "History of molecular beam research: Personal reminiscences of the important evolutionary period 1919–1933", *American Journal of Physics*, **43** (1975) 661–671,

but just as interesting is the story of the intellectual descendents of Stern and Gerlach's work. These descendants include lasers, atomic clocks and the global positioning system, magnetic resonance medical imaging, quantum computers, and a molecule that deactivates the AIDS virus. Some of this richness is described in

> Dudley R. Herschback, "Imaginary gardens with real toads", in *The Flight From Science and Reason*, edited by Paul R. Gross *et al.* (New York Academy of Sciences, New York, 1997) pages 19–24.

### 3.7   Problems

3.1   *Could friction account for these unexpected results?* Suppose that friction were important for atoms, so that after spending a short time in a magnetic field, all the atomic magnetic arrows would be pointing in the direction of the field. What results would you then expect from the Stern–Gerlach experiment?

3.2   *Real vs. ideal experiments.* How could Stern and Gerlach have known that the width of the peaks they observed was due only to the limits of their instrument and not to some property intrinsic to the atoms? Hint: Examine the figure below.

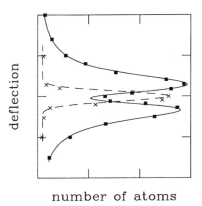

number of atoms

Fig. 3.4.   Results from a recent Stern–Gerlach experiment. Solid line and squares show results with non-uniform magnetic field, dashed line and crosses show results with no magnetic field. The vertical scale is magnified; the actual deflections span a range of less than two millimeters.

# 4

# The Conundrum of Projections;
# Repeated Measurements

## 4.1   The conundrum of projections

Whenever Stern and Gerlach measured the projection of a silver atom's
magnetic arrow on an axis, they found either $+m_B$ or $-m_B$. But the
figure below demonstrates that it is impossible for any arrow to have a
projection of $\pm m_B$ on *all* axes! Even if the projection onto the vertical
axis (in the figure, axis #1) happens to be $+m_B$, then we can always draw
some other axis (such as axis #2) that has a different projection (in the
figure, something more than $m_B$).

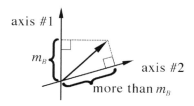

I call this difficulty the "conundrum of projections". The fact that the
projections can only take on the values of $\pm m_B$ is strange and unexpected,
but it's something that we can live with. (After all, much of human
behavior — and most of politics — is strange and unexpected too, once
you think about it.) The conundrum of projections is far more serious,
because it seems at first to be not just strange, but logically impossible.
In order to resolve the conundrum, we will introduce experiments in
which we actually measure the projection on various axes, and we will let
the results of those experiments suggest a resolution. Before doing this,
however, we must introduce some terminology.

21

*Terms for projections*

The figure below shows four different axes. In this book, except for section 11.1, we will consider only projections onto axes lying within the $(x, z)$ plane.

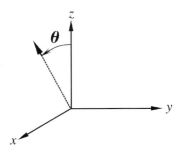

The projection of the arrow onto a vertical axis is called $m_z$.

The projection of the arrow onto a horizontal axis is called $m_x$.

The projection of the arrow onto a downward vertical axis is called $m_{(-z)}$.

The projection of the arrow onto an axis within the $(x, z)$ plane but tilted at an angle $\theta$ to the vertical is called $m_\theta$. (Thus $m_z = m_{0°}$, $m_x = m_{90°}$, and $m_{(-z)} = m_{180°}$.)

Note that if the projection onto some axis is $+m_B$, then the projection onto an axis pointing in the opposite direction is $-m_B$.

*Stern–Gerlach analyzer*

For convenience, I will package the Stern–Gerlach apparatus into a tall thin box and call it a *Stern–Gerlach analyzer*. There are only two places where a silver atom can come out of the apparatus, so the analyzer box has only two exit ports. (In the rest of this book, I will use only silver atoms and I will usually call them just "atoms" rather than "silver atoms".) The box also contains plumbing to the right of the non-uniform magnetic field which pushes the atoms around so that an outgoing atom follows a track parallel to the track of an incoming atom.* This plumbing doesn't affect the atom's magnetic arrow. These alterations do not change any important property of the Stern–Gerlach apparatus; they merely make our diagrams easier to read.

---

* One way to produce such plumbing is by installing a second non-uniform magnetic field that points in the opposite direction from the first.

In summary, the raw apparatus shown here:

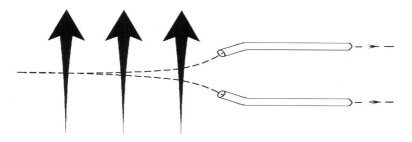

is packaged into a box and represented as:

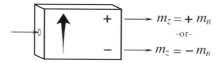

An atom enters the box on the left, and then either it leaves through the upper exit, marked +, in which case it has $m_z = +m_B$, or else it leaves through the lower exit, marked −, in which case it has $m_z = -m_B$.

On the other hand, if the Stern–Gerlach analyzer were oriented horizontally, then the exiting atom would have either $m_x = +m_B$ or else $m_x = -m_B$. Or, we could tilt the Stern–Gerlach analyzer box 17° to the right of vertical, in which case exiting atoms would have either $m_{17°} = +m_B$ or else $m_{17°} = -m_B$. In other words, a vertical analyzer measures $m_z$, a horizontal analyzer measures $m_x$, and our tilted analyzer measures $m_{17°}$.

## 4.2  Repeated measurement experiments

*Experiment 4.1.* Measurement of $m_z$, then $m_z$ again.

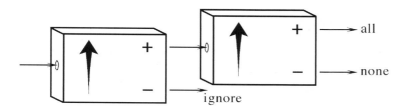

An atom entering the first analyzer will leave either the top (+) exit or the bottom (−) exit. In the latter case the exiting atom has $m_z = -m_B$ and we ignore it. In the former case the exiting atom has $m_z = +m_B$ and it is fed into the second analyzer. All such atoms leave the + exit of the second analyzer. In short, if an atom is found to have $m_z = +m_B$ at the first analyzer, then it does at the second analyzer as well.

*Experiment 4.2.* Measurement of $m_z$, then $m_{(-z)}$.

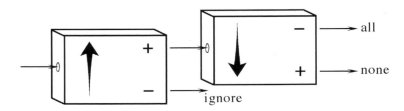

If an atom is found to have $m_z = +m_B$ at the first analyzer, then it has $m_{(-z)} = -m_B$ at the second analyzer.

*Experiment 4.3.* (The crucial experiment.) Measurement of $m_z$, then $m_x$, then $m_z$.

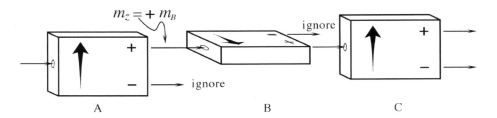

An atom entering analyzer A will leave from either the + exit or the − exit. In the latter case we ignore it, and in the former case we feed the atom (with $m_z = +m_B$) into analyzer B, a horizontal Stern–Gerlach analyzer that measures $m_x$. The atom will then either leave the − exit (in which case it has $m_x = -m_B$) and we ignore it, or else it will leave the + exit (in which case it has $m_x = +m_B$) and we feed it into analyzer C, a vertical analyzer that measures $m_z$.

What do you think will happen then? You might reason that this atom is known to have $m_z = +m_B$, because it left the + exit of analyzer A (as well as $m_x = +m_B$, because it left the + exit of analyzer B) and thus that it will leave the + exit of analyzer C, just as the atoms in experiment 4.1 did. This seems reasonable, and I will call it the "good guess argument". But in fact this *does not* happen. Instead, some atoms at this stage leave the + exit of analyzer C and others leave the − exit.

In summary, when an atom enters analyzer B it has a definite value of $m_z$, namely $+m_B$ — we know this because of experiment 4.1. But when that atom leaves analyzer B it *does not* have a definite value of $m_z$ — we know this because when it enters analyzer C it might leave through either the + or the − exit.

It's worth investigating this unexpected result further. We perform the experiment many times, and each time record whether the atom entering analyzer C leaves through the + exit or through the − exit. We find that there is no regular pattern to the exits, but that about half of the atoms leave through + and the rest leave through −. Thus although we cannot say with certainty which way the atom will leave analyzer C, we can say that it has probability one-half of leaving through either exit.

The following picture helps some people. They think of an atom leaving the + exit of analyzer B as having a magnetic arrow that points straight out of the page (that is, in the $+x$ direction). In classical mechanics, if such an atom entered analyzer C it would pass straight through. But the Stern–Gerlach result shows that in truth (that is, in quantum mechanics) it can't pass straight through — it must go either up or down (that is, it must leave through either the + exit or the − exit). If you "want" to go straight but are forced to go either up or down, the best you can do is go up half the time and down half the time. This picture is not entirely accurate (as we will see in detail later) but if you keep in mind both the picture and its limitations it may help you visualize the process.

I want to go back for a moment to the good guess argument, the one which suggests that every atom should leave analyzer C through the + exit. Experiment shows that this result is not correct, but we can also produce reasoning showing that it is not correct: We know that an atom leaving the + exit of analyzer B has $m_x = +m_B$. The good guess argument supposes that, by virtue of having previously left the + exit of analyzer A, it also has $m_z = +m_B$. You can see from the diagram below that an atom with both $m_x = +m_B$ and $m_z = +m_B$ would have a value for $m_{45°}$ that is *bigger* than $m_B$. (Experts in geometry will recognize from the diagram

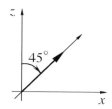

that in fact $m_{45°} = +\sqrt{2}m_B$, but you don't need to be an expert to see that $m_{45°}$ is larger than $m_B$.) But whenever $m_{45°}$ is measured, it is found to be either $+m_B$ or $-m_B$, and never to be bigger than $+m_B$!

An atom with a definite value for both $m_x$ and $m_z$ would have values for other projections that are not $\pm m_B$, and such atomic states do not exist. The flaw in the good guess argument is not in its reasoning, but in its assumptions. It assumed that an atom leaving analyzer B would have the same value of $m_z$ as it did when it entered, and this is false.

### 4.3    The upshot

We escape from the conundrum of projections via probability. If an atom has a definite value of the projection of its magnetic arrow on one axis, then it does *not* have a definite value of the projection of its arrow on some other axis. Given an atom with $m_x = +m_B$, to ask "What is the value of $m_z$?" is just like asking "What is the color of love?". These questions have no answers because for this atom, $m_z$ *doesn't have* a value in just the same way that love doesn't have a color. What *can* be said of such an atom is the probability of finding either of the two possible projections on the vertical axis.

Terminology note: Be wary of the phrase "definite value". When I say "An atom with a definite value of $m_x$ doesn't have a definite value of $m_z$" what I really mean is "An atom with a value of $m_x$ doesn't have a value of $m_z$". The second wording is more accurate and more clearly points out the difference between the quantal world and the classical world. But it is so stark that it makes most physicists uncomfortable. (It certainly makes *me* uncomfortable.) So usually I will employ the euphemism of "definite value". This is a personal failing of mine but I can't help it.

If the second analyzer were tilted at an angle $\theta$ relative to the first, then what would be the probability that an atom leaving the + exit of the first analyzer will leave the + exit of the second? We have so far discussed the situations $\theta = 0°$, $90°$, and $180°$ (in experiments 4.1, 4.3, and 4.2 respectively). In these situations the answers were $1$, $\frac{1}{2}$, and $0$. The experimentally determined answer to the question for any value of $\theta$ is given in figure 4.1. Notice that the graph interpolates smoothly between the known results at $\theta = 0°$, $90°$, and $180°$. For example, at $\theta = 60°$ the probability is $\frac{3}{4}$. (Experts in trigonometry will have already guessed the truth, namely that the probability is given by $\cos^2(\theta/2)$.)

### 4.4    Barriers to understanding

We have already reached the first central concept of quantum mechanics: *The outcome of an experiment cannot, in general, be predicted exactly; only the probabilities of the various outcomes can be found.* Many learners find their grasp beginning to slip at this point. If you are one of them, then don't flounder, but instead look inside of yourself to find the reason.

Is it that you hate math, so when I wrote down $\cos^2(\theta/2)$ you felt nauseous? Then relax: you'll never need to calculate with cosines.

Is it because you don't care about magnetic needles and don't want to learn about them? Then remember that I'm using magnetic needles only as an example to illustrate the principles of quantum mechanics, and that those principles describe all the actions in the universe.

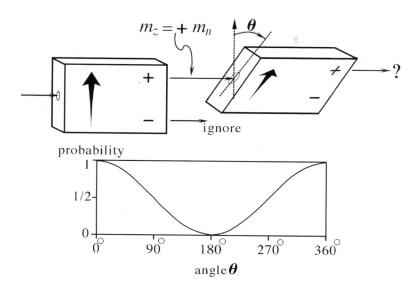

Fig. 4.1. The probability of an atom leaving the + exit of the second analyzer as the tilt angle $\theta$ between the two analyzers is varied. The probability is 1 for $\theta = 0°$, $\frac{3}{4}$ for $\theta = 60°$, $\frac{1}{2}$ for $\theta = 90°$, $\frac{1}{4}$ for $\theta = 120°$, 0 for $\theta = 180°$, etc.

A deeper problem bothers those who say "I see *that* only probabilities can be found, but I want to know *why* only probabilities can be found." Fundamentally, I have no answer to this concern. I don't know why the universe works the way it does any more than you do. I'm not God, I didn't create the universe, so don't complain to me. However, I suspect that when you ask the question "why?", you're really worried about something else. There are lots of good "why" questions that you never ask: Why does the universe have three dimensions? Why do we eat pancakes often for breakfast but rarely for dinner? Why do women wear skirts and men pants? You don't ask these questions because you're so familiar with the facts that you never stop to question why they're true. I think that most people who ask "Why can only probabilities be found?" are really just crying out that the new world of quantum mechanics is strange and unfamiliar. It certainly is. But this should be seen as a challenge to invite exploration rather than an excuse to crawl back into your familiar, secure, classical hole.

Finally, the most dangerous barrier to understanding of all: You don't want the result to be true. It seems strange — it *is* strange — so you simply reject it. But all sorts of things seem strange upon first encounter. When it was first discovered that the earth was round, that must have seemed strange too! I admit that even though I have studied a lot of quantum mechanics, it still seems strange to me, but it seems strange and delightfully quirky, rather than strange and repulsive. If you are rejecting quantum mechanics simply because it's strange, then I urge you to keep at it until you find it as beautiful as I do.

## 4.5   Sample problem

In experiment 4.3, half the atoms entering analyzer C leave through the + exit and half leave through the − exit. Suppose the experiment is altered by tilting analyzer A 30° to the right of vertical. Analyzers B and C are not changed. In this new experiment, what portion of the atoms entering analyzer C will leave through the + exit?

### Solution

An atom leaving the + exit of analyzer B has $m_x = +m_B$. It doesn't care what state it was in when it entered analyzer B — it could have come directly from an oven, or it could have come through a complicated set of a dozen analyzers tilted at various angles — the output state is specified completely by saying $m_x = +m_B$. Thus half of the atoms entering analyzer C will leave through the + exit whether analyzer A is tilted to 30°, 0°, or any other angle.

## 4.6   Problems

4.1   *The conundrum of projections.* An arrow is three inches long and points due west. What is its projection on an axis that points: (a) due west, (b) due east, (c) due north, (d) straight up, (e) straight down, (f) half-way between straight up and due west?

4.2   *Two analyzers.* In experiment 4.1 on page 23 the atoms leaving the − exit of the first analyzer were ignored. What would happen to them if they were instead fed into another vertical analyzer?

4.3   *Certainty.* I have claimed that "the outcome of an experiment cannot, in general, be predicted exactly; only the probabilities of the various outcomes can be found". Yet in experiment 4.1 on page 23 an atom entering the second analyzer will certainly leave through the + exit. How can my claim and this experiment be reconciled?

4.4   *The state of an atom.* Which phrase best describes the state of an atom that leaves the + exit of analyzer B in experiment 4.3 on page 24: (1) It has $m_z = +m_B$. (2) It has $m_x = +m_B$. (3) It has both $m_z = +m_B$ and $m_x = +m_B$.

4.5   *Three analyzers.* In experiment 4.3 on page 24 the atoms leaving the — exit of analyzer B were ignored. What would happen to them if they were instead fed into another horizontal analyzer? Into a vertical analyzer?

4.6   *Three analyzers rearranged.* In experiment 4.3 on page 24, atoms leave the three analyzers according to statistics described in this table:

| analyzer | exit statistics |
|----------|-----------------|
| A | depends on character of incoming atoms |
| B | half through +, half through — |
| C | half through +, half through — |

If analyzer A were lifted so that the atoms entering analyzer B came from the — exit of A (rather than from the + exit), how would the table change?

4.7   *Rotations.* Would any of the results in this chapter change if the entire experimental apparatus (source, all analyzers, and detectors) were rotated as a unit?

4.8   *Different angles.* A careful reading of the graph in figure 4.1 shows that if the first analyzer is vertical, and the second is tilted to the right of vertical by 60°, then the probability of an atom leaving the + exit of the second analyzer is $\frac{3}{4}$. What would be the probability if the first analyzer were 60° to the *left* of vertical and the second were vertical?

4.9   *More different angles.* Two Stern–Gerlach analyzers are arranged as shown below. Analyzer A is tilted 45° to the left of vertical, while analyzer B is tilted 45° to the right of vertical. Atoms leaving the + exit of A are fed into the input of B. What is the probability that an atom entering B will leave it through the + exit?

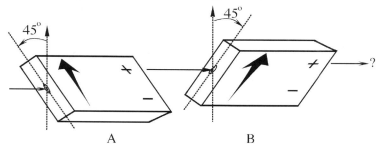

Hint: What would happen if A were tilted 10° to the left, and B were tilted 80° to the right?

4.10  *Three analyzers with different angles.* Consider experiment 4.3 on page 24, but suppose that analyzer B were not horizontal, but rather tilted to the right of vertical by 60°. In this case, what is the probability that an atom entering analyzer C will emerge from the + exit? From the − exit? Hint: See figure 4.1.

4.11  *Barriers to understanding.* Distinguish between "a description of the rules of chess", "an understanding of the rules of chess", and "an explanation for the rules of chess". Which of these do you need to play a good game of chess?

4.12  *Familiar vs. understood.* My mother once told me that "I used to understand telephones, but I don't understand these new cellular phones." When I asked her how a conventional telephone worked, she could only say "I think it has carbon in it." In three or fewer sentences, show how this story illustrates the difference between familiarity and understanding. (My mother is, by the way, perfectly capable of using any sort of telephone.)

4.13  *Explaining Newtonian and quantum mechanics.* (For technical readers.)

(a) In Newtonian mechanics, force is related to acceleration ($\mathbf{F} = m\mathbf{a}$), whereas most laymen believe that force is related to speed. (Lay belief: "If you push something, it moves." Newton: "If you push something, its motion changes.") How would you respond to an intelligent layman who asked you why Newtonian mechanics is correct?

(b) In classical mechanics, the future can be predicted exactly, whereas in quantum mechanics only probabilities can be found. How would you respond to an intelligent layman who asked you why quantum mechanics is correct?

# 5

# Probability

I interpreted the repeated measurement experiments of the previous chapter by saying that quantum mechanics can find probabilities only, not certainties (that is, that quantum mechanics is "probabilistic", not "deterministic"). You may object, maintaining that the world is deterministic, but that my particular deterministic scheme (the "magnetic arrow") is incorrect. The next chapter presents an ingenious argument, invented by Einstein, which shows that *no* local deterministic scheme could give the results observed by experiment. In order to understand that argument you need some background in probability.

But in fact, a knowledge of probability is generally useful in day-to-day life as well as in physics. You walk across a street — what is the probability of your being hit by a car? You are advised to undergo elective surgery — what is the probability that the surgery will extend your life, and what is the probability that the surgery will go wrong and injure you? You breath some asbestos or smoke a cigarette — what is the probability of contracting cancer? Misconceptions about probability abound and can lead to disastrous public policy decisions.* A knowledge of quantum mechanics is good for your soul, but it is of practical importance only to the designers of lasers, transistors, and superconductors. A knowledge of probability is of practical importance to everyone.

---

* Suppose that, in a democratic society, 70% of the citizens prefer to drink beer and 30% prefer to view artwork. Does this imply ("majority rules") that all art museums should be converted into bars? Does it imply that the ratio of bars to art museums ought to be fixed by law at 7 to 3? Of course it implies neither. But many policy makers seem never to have learned this simple lesson in probability.

## 5.1   Gambling probability

If you toss a die, the probability of rolling a **2** is $\frac{1}{6}$. If you flip a coin, the probability of getting **heads** is $\frac{1}{2}$. In general, for gambling probabilities,

$$\text{probability of a success} = \frac{\text{number of successful outcomes}}{\text{number of possible outcomes}}.$$

This rule holds only for gambling probabilities like those we have just mentioned. It does not apply, for example, to surgery, where there are only two possible outcomes — survival and death — but the chance of survival is far greater than $\frac{1}{2}$. Nor does it apply to the Stern–Gerlach analyzer, where an incoming atom can leave through either the + exit or through the − exit, but figure 4.1 shows that the probabilities of these two possible outcomes are not always 50%. Finally, if you buy a lottery ticket, there are only two possible outcomes — winning and losing — but the probability of winning is sadly less than $\frac{1}{2}$.

## 5.2   Compound probabilities

*Example 1.* Toss a die. What is the probability of rolling either a **1** or a **3**? In this case, there are six possible outcomes and two successful outcomes, so the probability of success is $\frac{2}{6} = \frac{1}{6} + \frac{1}{6}$. In general, the word "or" is a signal to *add* probabilities.

*Example 2.* Toss a die and simultaneously flip a coin. What is the probability of getting **2** and **tails**? In this case there are twelve possible outcomes (**1** and **heads**, **1** and **tails**, **2** and **heads**, **2** and **tails**, and so forth up to **6** and **tails**) so the probability of getting **2** and **tails** is $\frac{1}{12} = \frac{1}{6} \times \frac{1}{2}$. In general, the word "and" is a signal to *multiply* probabilities.

*Example 3.* Flip three coins (or flip one coin three times):

| possible outcome | probability of this outcome | number of heads |
|---|---|---|
| HHH | 1/8 | 3 |
| HHT | 1/8 | 2 |
| HTH | 1/8 | 2 |
| THH | 1/8 | 2 |
| HTT | 1/8 | 1 |
| THT | 1/8 | 1 |
| TTH | 1/8 | 1 |
| TTT | 1/8 | 0 |

Thus the probability of obtaining three heads is $\frac{1}{8}$, of two heads is $\frac{3}{8}$, of one head is $\frac{3}{8}$, of no heads is $\frac{1}{8}$.

*Example 4.* Toss two dice (or toss one die two times). What is the probability that the sum of the face-up dots is four? If the first die lands on **4** and the second on **3**, I will call the outcome "[**4**, **3**]".

probability of [**1**, **1**] is $\frac{1}{6} \times \frac{1}{6} = \frac{1}{36}$
probability of [**1**, **2**] is $\frac{1}{6} \times \frac{1}{6} = \frac{1}{36}$
probability of [**2**, **1**] is $\frac{1}{6} \times \frac{1}{6} = \frac{1}{36}$
probability of [**1**, **3**] is $\frac{1}{6} \times \frac{1}{6} = \frac{1}{36}$
probability of [**3**, **1**] is $\frac{1}{6} \times \frac{1}{6} = \frac{1}{36}$
probability of [**2**, **2**] is $\frac{1}{6} \times \frac{1}{6} = \frac{1}{36}$
    and so forth to
probability of [**5**, **6**] is $\frac{1}{6} \times \frac{1}{6} = \frac{1}{36}$
probability of [**6**, **5**] is $\frac{1}{6} \times \frac{1}{6} = \frac{1}{36}$
probability of [**6**, **6**] is $\frac{1}{6} \times \frac{1}{6} = \frac{1}{36}$.

Now

probability that the sum tossed is four =
    (probability of [**2**, **2**]) +
    (probability of [**1**, **3**]) +
    (probability of [**3**, **1**]),

but

(probability of [**2**, **2**]) = (prob. of **2**) × (prob. of **2**) = $\frac{1}{6} \times \frac{1}{6}$,
(probability of [**1**, **3**]) = (prob. of **1**) × (prob. of **3**) = $\frac{1}{6} \times \frac{1}{6}$,
(probability of [**3**, **1**]) = (prob. of **3**) × (prob. of **1**) = $\frac{1}{6} \times \frac{1}{6}$,

thus

probability that the sum tossed is four =
$(\frac{1}{6} \times \frac{1}{6}) + (\frac{1}{6} \times \frac{1}{6}) + (\frac{1}{6} \times \frac{1}{6}) = \frac{1}{12}$.

## 5.3  Tilting Stern–Gerlach analyzer

Mount a single Stern–Gerlach analyzer on a pivot so that it can be tilted to have its magnetic field point in any of the three directions **A**, **B**, or **C**. In the figure on the next page, it is tilted to orientation **A** on the left and to orientation **B** on the right. This analyzer is switched at random between these three orientations, each orientation having probability $\frac{1}{3}$. Suppose an atom with $m_z = +m_B$ were fed into the analyzer. (For example, the atom might have just emerged from the + exit of an analyzer fixed in orientation **A**.) What is the probability that it leaves through the + exit?

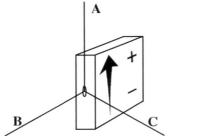 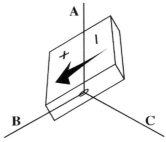

If the orientation is **A**,
   the probability that the atom leaves the + exit is 1.
If the orientation is **B**,
   the probability that the atom leaves the + exit is $\frac{1}{4}$.
If the orientation is **C**,
   the probability that the atom leaves the + exit is $\frac{1}{4}$.

(These last two facts come from figure 4.1 on page 27: when $\theta = 120°$ or $\theta = 240°$, the probability is $\frac{1}{4}$.)
Now

   probability that atom leaves from + exit =
      (probability that it does so when orientation is **A**) +
      (probability that it does so when orientation is **B**) +
      (probability that it does so when orientation is **C**)

but

   probability that it does so when orientation is **A** = $\frac{1}{3} \times 1$
   probability that it does so when orientation is **B** = $\frac{1}{3} \times \frac{1}{4}$
   probability that it does so when orientation is **C** = $\frac{1}{3} \times \frac{1}{4}$

thus

   The probability that an atom entering with $m_z = +m_B$
   leaves from the + exit is
      $(\frac{1}{3} \times 1) + (\frac{1}{3} \times \frac{1}{4}) + (\frac{1}{3} \times \frac{1}{4}) = \frac{1}{2}$.

## 5.4   References

   Warren Weaver, *Lady Luck: The Theory of Probability* (Doubleday, Garden City, New York, 1963).
   Vinay Ambegaokar, *Reasoning About Luck: Probability and Its Uses in Physics* (Cambridge University Press, New York, 1996).
   Harold Warren Lewis, *Why Flip a Coin? The Art and Science of Good Decisions* (John Wiley, New York, 1997).

## 5.5  Problems

It is important that you look at the first six problems in this section. The remaining problems are fun and informative but are not needed to support this book's train of argument.

5.1  *Three dice.* If you throw three dice, what is the probability that a total of four dots are face up?

5.2  *License plates.* Suppose that in the state of Iowa auto license plates are identified by three letters followed by three numbers, and that the numbers are chosen at random. If you glance at an Iowa plate, what is the probability that the two last digits will be the same?

5.3  *Two dice.* Throw two dice. What is the probability that the sum of the face-up dots is more than four?

5.4  *The coin toss.* Toss a single coin ten times.

   (a) What is the probability of obtaining all heads (the pattern HHHHHHHHHH)?

   (b) What is the probability of obtaining alternating heads then tails (the pattern HTHTHTHTHT)?

   (c) What is the probability of obtaining the pattern HTTTHHTTHT?

   (d) What is the probability of obtaining a pattern with one tail and nine heads?

5.5  *Tilting Stern–Gerlach analyzer.* In section 5.3 we found the probability for an atom that entered a tilting Stern–Gerlach analyzer with $m_z = +m_B$ to leave through the $+$ exit. What is that probability for an atom that enters:

   (a) With $m_z = -m_B$?

   (b) With $m_{120^\circ} = +m_B$?

   (c) With $m_{(-120^\circ)} = +m_B$?

5.6  *Military draft lottery.* From 1967 to 1972 the United States used a military draft lottery in which birthdays were selected at random, and the army drafted first men born on the first date selected, then men born on the second date selected, and so forth. Suppose a small country uses a similar system, but the country's records include not birth date, but only birth season: summer or winter. In the year 1968 there are 1000 draftable men of whom 600 were born in winter and 400 were born in summer. The military requires 700 draftees. (Thus a "fair" system would assign each man a probability 7/10 of being drafted.) The country holds a lottery to determine whether summer-born or winter-born men will be called up first (either possibility has probability 1/2). Within each birth category, the military drafts men at random. The drafting continues from the first category into the second until the requirement of 700 draftees is filled.

(a) Show that if the summer-born are called up first, the probability of a winter-born man being drafted is 1/2.

(b) Show that if the winter-born are called up first, the probability of a summer-born man being drafted is 1/4.

(c) What is the overall probability of drafting a summer-born man? A winter-born man? (The term "overall probability" means the probability that would be calculated *before* the lottery was held.)

Comment upon the fairness of this draft lottery scheme.

5.7 *Speeding tickets.* Benjamin Marrison of the Cleveland *Plain Dealer* wondered whether Ohio state troopers were more likely to hand out speeding tickets at the end of the month ("Spotting the speeders: How, when, and where troopers will get you", 3 December 1995). He uncovered data showing that for the year 1994, a total of 19 737 speeding tickets were issued on the 28th of some month, 19 623 were issued on the 30th, but only 18 845 were issued on the 31st. From this he concluded that one is actually less likely to get a speeding ticket at the end of a month. Comment.

5.8 *Outdoor hazards.* From the New York *Times* (30 June 1998): "Where are you more likely to be injured: climbing down a rock face or sitting around a campsite? ... A new study of the hazards of national parks found that injuries at campsites ... outnumber those sustained during rock climbing by more than 3 to 1." Comment.

5.9 *Winning the lottery twice.* Suppose that the state of New Jersey runs one lottery game each week, that each week there is one and only one winner, that five million individuals play each week, that each of those individuals buys a single ticket, and that this same group of individuals plays every week. (These suppositions are, of course, not precisely correct. On the other hand they are not terribly different from the actual situation, and this simplification allows the situation to be understood much more readily than a "perfect portrait" would permit.) What is the probability that:

(a) One particular player, Sylvia Struthers of Clinton Mills, New Jersey, wins in the two successive lotteries held on the ninth and tenth weeks of the year 1998.

(b) A player, not necessarily Ms. Struthers, wins in both of these two lotteries.

(c) A player wins in successive weeks in 1998.

(d) A player wins twice in 1998.

5.10 *Average vs. typical, part 1.* A politician claims that she "always stands up for the average man". Does that mean that she always supports the majority of people? Hint: How many people do you know who are of average height?

5.11 *Average vs. typical, part 2.* At the Lincoln Street branch of the First National Bank, a customer needing a teller enters the bank, on average, once each minute, and each teller transaction lasts, on average, three minutes. Based on these facts, the branch manager decides to staff the bank with exactly three tellers. Why was the branch manager fired?

5.12 *Correlations.* In a poll, one thousand individuals are asked about their preferences in music and in dining. One-tenth of the individuals preferred opera to rock, and one-fifth of them preferred French restaurants to fast food restaurants. Is the probability that one of the polled individual prefers *both* opera *and* French restaurants equal to 2% ? ($2\% = \frac{1}{50} = \frac{1}{10} \times \frac{1}{5}$.)

5.13 *Random vs. haphazard.* Smith and Jones are running for congress, and Ms. Struthers wants to know who will probably win. So she polls ten of her friends, and finds that eight plan to vote for Smith and two for Jones. She is surprised and shocked when Jones wins by a margin of 54% to 46%. What was wrong with her poll?

# 6

# The Einstein–Podolsky–Rosen Paradox

My interpretation of the repeated measurement experiments in section 4.2 was:

> An atom with a definite value of $m_z$ doesn't have a definite value of $m_x$. All that can be said is that when $m_x$ is measured, there is probability $\frac{1}{2}$ of finding $+m_B$ and probability $\frac{1}{2}$ of finding $-m_B$.

This is in many ways the simplest and most natural interpretation, but there are other possibilities. For example, the "measurement disturbs a classical system" possibility:

> An atom with a definite value of $m_z$ also has a definite value of $m_x$, but the measurement of $m_z$ disturbs the value of $m_x$ in an unpredictable way.

or the "complex atom" possibility:

> An atom with a definite value of $m_z$ also has a definite value of $m_x$, but this value changes so rapidly that no one can figure out what that value is.

The Einstein–Podolsky–Rosen (or EPR) argument shows that both of these "other interpretations" are untenable.

I will give the argument in the form of two hypothetical experiments. Because of technical difficulties, these experiments have never been carried out in exactly the form that I will describe. But similar experiments have been performed, most notably by Alain Aspect and his collaborators at the University of Paris's Institute of Theoretical and Applied Optics at Orsay. Figure 6.1 shows the apparatus that this group employed and, as usual, it is much more elaborate than the sketch diagrams that I will use later to describe the hypothetical experiment. Our hypothetical experiment

38

Fig. 6.1.   Alain Aspect's laboratory in Orsay, France (courtesy of A. Aspect).

will employ a pair of atoms and detectors that tilt by 120°. Aspect's real experiment employed a pair of "photons" ("particles of light") and detectors that tilted by 22.5°. In spite of these technical differences, the real experiment was *conceptually* equivalent to the one I will describe here, and its results are a ringing endorsement of quantum mechanics.

## *Locality*

Before proceeding, I must attend to one small but essential point: the term "local". It is clear that something which happens at one place can influence what happens far away. For example, a newspaper article printed in Madrid can foment a revolution in Buenos Aires. But the effect happens some time after the cause, because it takes some time for the agent of influence (the newspapers) to travel from Madrid to Buenos Aires, and as they travel they always move bit by bit — they never disappear from one place and reappear at another without passing through intermediate points. This method of influence is called "local". Modern communication technology might appear to be non-local, because when you speak into a telephone it seems that you can be heard far away at the same instant. But in fact there is a short — and usually unnoticeable — delay between the speaking and the hearing, as electrical signals encoding your voice travel through telephone lines at the speed of light.

> *Technical aside:* Notice that the very definition of locality involves concepts like cause and effect, concepts that assume a deterministic world. Because quantum mechanics is not deterministic and events can take place without causes, the concept

of locality becomes more subtle and complex. The technical literature is thus full of terms like "active locality", "passive locality", "non-locality", and "alocality".

The assumption of locality is so natural and commonplace that it has been enshrined in poetry:[*]

> And when the loss has been disclosed, the Secret Service say:
> "It *must* have been Macavity!" — but he's a mile away.

Einstein's theory of relativity puts the assumption of locality on an even firmer basis, establishing that no causal agent can travel faster than a light signal. Standard quantum mechanics, as presented in this book, retains the assumption of locality. But it *is* possible to produce alternatives to standard quantum theory that are non-local.

I mention locality here because the experiments described below illuminate our old ideas in a strange — but ultimately satisfying — new light.

### 6.1    Experiment 6.1: Distant measurements

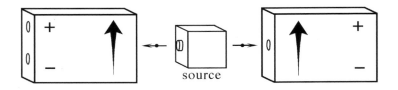

In this experiment a box labeled "source" produces a pair of atoms with a net magnetic arrow of zero, and the two atoms fly off in opposite directions. Each atom is detected by its own vertical Stern–Gerlach analyzer.

Observed results: The probability that the right atom leaves through the + exit is $\frac{1}{2}$, the probability that it leaves through the − exit is $\frac{1}{2}$. Similarly for the left atom. But if the right atom leaves through its + exit, then the left atom always leaves through its − exit, and vice versa. This is true regardless of which, if either, analyzer is closer to the source. It is also true regardless of the orientation of the two analyzers, as long as both have the same orientation.

Here is a straightforward proposal that explains most of these observations: Simply suppose that when the pairs of atoms are produced, one atom has $m_z = +m_B$ and the other has $m_z = -m_B$. This proposal explains the first four observations but it is inconsistent with the last one. If the two analyzers are, say, horizontal instead of vertical, then under this proposal

---
[*] T.S. Eliot, *Old Possum's Book of Practical Cats*.

it would be possible (see problem 6.2) for both atoms to leave through their + exits, or for both to leave through their − exits. But in fact the two atoms always leave through opposite exits. The straightforward proposal is appealing, but it must be wrong. Eventually we will replace the straightforward yet incorrect proposal with a much more elaborate one, a proposal called "quantum mechanics". For the time being, however, it is important to get a clear idea of how atoms *actually do* behave before rushing into new proposals. So how do atoms behave?

Imagine, for example, that the left analyzer is five miles from the source, while the right analyzer is five miles plus one inch from the source. Then the left atom will go into its analyzer and be measured before the right atom goes into its analyzer. Suppose that the left atom leaves the + exit. Then it is known with certainty that the right atom has $m_z = -m_B$ (i.e. that when it gets to its analyzer it will leave through the − exit, but the right atom itself has not been measured. It is impossible that the right atom, ten miles away from the scene of the measurement, could have been mechanically disturbed by the measurement of the left atom.[†] The first alternative interpretation mentioned on page 38 must be rejected.

If you are familiar with Einstein's theory of relativity, you know that the fastest possible speed at which a message can travel is the speed of light. Yet this experiment suggests a mechanism for instantaneous communication: When the two atoms are launched, it cannot be predicted whether the right atom will leave the + exit or the − exit once it gets to its analyzer. But the instant that the left atom leaves the + exit of its analyzer, it is known that the right atom (now ten miles away) will leave the − exit once it gets to the right analyzer. This seems to be instantaneous communication. But the important point is not whether "it is known that the right atom will leave the − exit" but rather *who* knows that the right atom will leave the − exit. Certainly the person standing next to the left-hand analyzer knows it[‡] but the person on the left won't be able to tell the person on the right except through some ordinary, slower-than-light mechanism. The result is strange (Einstein called it "spooky") but it does not open up the door to instantaneous communication.

Quantum mechanics forces us to the brink of implausibility — *but not beyond.*

> *Technical aside:* The conceptual equivalent of this experiment
> has been performed many times, usually with detectors located
> yards rather than miles apart. But in 1997 Nicolas Gisin of

---

[†] Is it really "impossible"? In fact, this is the assumption of locality which, as I have mentioned, is very natural but nevertheless an assumption.

[‡] And the person standing next to the right-hand analyzer knows that the person standing next to the left-hand analyzer knows it.

the University of Geneva and his collaborators performed the experiment with detectors in the Swiss villages of Bellevue and Bernex, separated by nearly seven miles.

## 6.2    Experiment 6.2: Random distant measurements

This experiment is called the "test of Bell's theorem". The reasoning is intricate, so I give an outline here before plunging into the details. We will build an apparatus much like the previous one with a central source that produces a pair of atoms, and with two detector boxes. Mounted atop each detector box are a red lamp and a green lamp. Every time the experiment is run, a single lamp on each detector box lights up. On some runs the detector on the left flashes red and the detector on the right flashes green, on other runs both detectors flash red, etc. When the apparatus is analyzed by quantum mechanics, we find that the probability of each detector flashing a different color is $\frac{1}{2}$. But we can also analyze the apparatus under the assumption of local determinism. This analysis shows that the probability of each detector flashing a different color is $\frac{5}{9}$ or more. (Exactly how much more depends on exactly which local deterministic scheme is employed, see problem 6.4.) Experiment agrees with quantum mechanics, so the assumption of local determinism, natural though it may be, is false. Any local deterministic scheme, including the second alternative interpretation mentioned on page 38, must be wrong.

### *The apparatus*

This experiment uses the same source as the previous experiment, but now the detectors are not regular Stern–Gerlach analyzers, but the tilting Stern–Gerlach analyzers described in section 5.3 (page 33). Each of the two analyzers has probability $\frac{1}{3}$ of being oriented as **A**, **B**, or **C**. If you wish, you may set the detector orientations and then have the source generate its pair of atoms, but you will get the same results if you first launch the two atoms and then set the detector orientations while the atoms are in flight. Mounted on each detector are two colored lamps. If an atom comes out of the + exit, the red lamp flashes; if an atom comes out the − exit, the green lamp flashes.

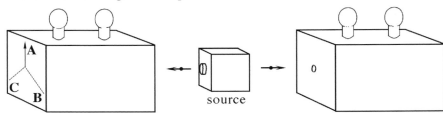

## The prediction of quantum mechanics

If the two detectors happen to have the same orientation, then this experiment is exactly the same as the previous one, so exactly the same results are obtained: the two detectors always flash different colors. On the other hand, if the two detectors have different orientations, then they might or might not flash different colors.

What is the probability that the two detectors flash different colors in general, that is, when the two detectors might or might not have the same orientation? Suppose the detector on the left is closer to the source than the detector on the right. If the left detector were set to **A** and flashed green (that is, −), then the atom on the right has $m_z = +m_B$. In the previous chapter we saw that when such an atom enters the right detector, it has probability $\frac{1}{2}$ of causing a red flash and probability $\frac{1}{2}$ of causing a green flash. You can readily generalize this reasoning to show that regardless of orientation, the two detectors flash different colors with probability $\frac{1}{2}$.

We conclude that:

(1) If the orientation settings are the same, then the two detectors flash different colors always.
(2) If the orientation settings are ignored, then the two detectors flash different colors with probability $\frac{1}{2}$.

And these results are indeed observed!

## The prediction of local determinism

In any local deterministic scheme, each atom must leave the source already supplied with an instruction set that determines which lamp flashes for each of the three orientation settings. For example, an instruction set might read (if set to **A** then flash red, if set to **B** then flash red, if set to **C** then flash green), which we abbreviate as (RRG). One natural way to implement an instruction set scheme would be through the atom's associated magnetic arrow: if the detector is vertical (orientation **A**) and the atom's arrow points anywhere north of the equator, then the atom leaves through the + exit, while if the atom's arrow points anywhere south of the equator, then the atom leaves through the − exit. Similar rules hold for orientations **B** and **C**: the atom always leaves through the exit towards which its arrow most closely points.[§] The argument that follows holds for

---

[§] This postulated scheme is inconsistent with quantum mechanics because it assumes that an atom's magnetic arrow points in the same manner that a classical stick does, with definite values for all three projections $m_x$, $m_y$, and $m_z$ simultaneously.

this natural scheme, but it also holds for any other oddball instruction set scheme as well.

To explain observation (1) above, assume that the two atoms are launched with opposite instruction sets: if the atom going left is (GRG), then the atom going right is (RGR), and so forth. (In the "natural" scheme, the two atoms are launched with magnetic arrows pointing in opposite directions.) Now let's see how we can explain observation (2).

If the instruction set for the atom going left is (RRG), and for the atom going right is (GGR), then what colors will the detectors flash? That depends on the orientation settings of the two detectors. Suppose the left detector were set to **C** and the right detector were set to **A**. Then the third letter of (RRG) tells us that the left detector would flash green, and the first letter of (GGR) tells us that the right detector would flash green. The same list-lookup reasoning can be applied to any possible orientation setting to produce the following table.

| orientation settings | detectors flash |
|:---:|:---|
| **AA** | RG: different |
| **BB** | RG: different |
| **CC** | GR: different |
| **AB** | RG: different |
| **BA** | RG: different |
| **BC** | RR: same |
| **CB** | GG: same |
| **AC** | RR: same |
| **CA** | GG: same |

There are nine possible orientation settings and five of them lead to different color flashes. So if the atom going left is (RRG), then the probability of different color flashes is $\frac{5}{9}$. A little thought shows that the same result applies if the atom going left is (GGR), or (GRG), or anything but (RRR) and (GGG). In the last two cases, the probability of different color flashes is of course 1.

Now we know the probability of different color flashes for any given instruction set. We want to find the probability of different color flashes period. To calculate this we need to know what kind of atoms the source makes. (If it makes only (RRR)s paired with (GGG)s then the probability of different color flashes is 1. If it makes only (RRG)s paired with (GGR)s then the probability of different color flashes is $\frac{5}{9}$. If it makes [(RRR) paired with (GGG)] half the time and [(RRG) paired with (GGR)] half the time, then the probability of different color flashes is half-way between $\frac{5}{9}$ and 1.)    Because I don't know exactly how the source works, I can't say exactly what the probability for different color flashes is.    But I do know that any source can make only eight kinds of atoms, because only eight kinds of atoms exist:

| kind of atom going left | probability of different color flashes |
|---|---|
| (RRR) | 1 |
| (GGG) | 1 |
| (RRG) | 5/9 |
| (RGR) | 5/9 |
| (GRR) | 5/9 |
| (RGG) | 5/9 |
| (GRG) | 5/9 |
| (GGR) | 5/9 |

Thus for any kind of source, the probability of different color flashes is some mixture of probability 1 and probability $\frac{5}{9}$.

We conclude that in any instruction set scheme, the detectors will flash different colors with probability $\frac{5}{9}$ (55.5%) *or more*.

### The conclusion

But in fact, the detectors flash different colors with probability $\frac{1}{2}$! The assumption of local determinism has produced a conclusion which is violated in the real world, and hence it must be wrong. Probability is not just the *easiest* way out of the conundrum of projections, it is the *only* way out.

> *Technical aside:* What, only? Well, almost only. In fact, our arguments only rule out the existence of instruction sets, and hence it permits alternatives to standard probabilistic quantum theory that do not rely on instruction sets. David Bohm, and others, have invented such deterministic but non-local alternatives. If you dislike quantum mechanics because it's too weird for your tastes, this may make you happy.   However, these alternative theories are necessarily pretty weird themselves. For example, in Bohm's theory the two atoms don't need instruction sets because they can communicate with each other instantaneously. To be absolutely accurate, probability is the only *local* way out of the conundrum of projections.

## 6.3    References

The subject of this chapter has a rich intellectual heritage. The general idea was introduced in

> A. Einstein, B. Podolsky, and N. Rosen, "Can quantum-mechanical description of physical reality be considered complete?", *Physical Review*, **47** (1935) 777–780,

and the specific form of our experiment 6.1 was devised by

> David Bohm, *Quantum Theory* (Prentice-Hall, Englewood Cliffs, New Jersey, 1951) pages 611–623.

But the most powerful part of the argument, the one embodied in our experiment 6.2, was developed by John Bell in 1964 and is called "Bell's theorem". Bell's writings on quantum mechanics, ranging from the popular to the very technical, are collected in

> John S. Bell, *Speakable and Unspeakable in Quantum Mechanics* (Cambridge University Press, Cambridge, UK, 1987).

The best semi-popular treatment of Bell's theorem and the Einstein–Podolsky–Rosen paradox (and the inspiration for much of this chapter) is

> N.D. Mermin "Is the moon there when nobody looks? Reality and the quantum theory", *Physics Today*, **38** (4) (April 1985) 38–47; see also the letters reacting to this article: *Physics Today*, **38** (11) (November 1985) 9–15, 136–142.

A computer program to simulate this test of Bell's theorem is

> Darrel J. Conway, *BellBox* (Physics Academic Software, Raleigh, North Carolina, 1993).

Real experimental results mentioned in this chapter were reported in

> Alain Aspect, Philippe Grangier, and Gérard Roger, "Experimental realization of Einstein–Podolsky–Rosen–Bohm *gedankenexperiment*: A new violation of Bell's inequalities", *Physical Review Letters*, **49** (1982) 91–94,
>
> Alain Aspect, Jean Dalibard, and Gérard Roger, "Experimental test of Bell's inequalities using time-varying analyzers", *Physical Review Letters*, **49** (1982) 1804–1807,
>
> W. Tittel, J. Brendel, B. Gisin, T. Herzog, H. Zbinden, and N. Gisin, "Experimental demonstration of quantum correlations over more than 10 km", *Physical Review* A, **57** (1998) 3229–3232,

but these papers are difficult for non-physicists to read. You might want to look instead at the reviews

> Arthur L. Robinson, "Quantum mechanics passes another test", *Science*, **217** (30 July 1982) 435–436,
> Arthur L. Robinson, "Loophole closed in quantum mechanics test", *Science*, **219** (7 January 1983) 40–41,
> Andrew Watson, "Quantum spookiness wins, Einstein loses in photon test", *Science*, **277** (25 July 1997) 481.

A recent high-accuracy test of Bell's theorem is described in

> P.G. Kwiat, E. Waks, A.G. White, I. Appelbaum, and P.H. Eberhard, "Ultra-bright source of polarization-entangled photons", *Physical Review* A, **60** (August 1999) R773–R776.

## 6.4 Problems

6.1 *Instantaneous communication.* In your own words, explain why you cannot send a message instantaneously using the mechanism of experiment 6.1. If quantum mechanics were deterministic rather than probabilistic, yet the distant atoms still always left from opposite exits, would you then be able to send a message instantaneously? What if the operator of the left-hand Stern–Gerlach analyzer were somehow¶ able to force his atom to come out of the + exit? (You might want to answer by completing the following story: "An eccentric gentleman in London has two correspondents: Ivan in Seattle and Veronica in Johannesburg. Every Monday he sends each correspondent a letter, and the two letters are identical except that he signs one in red ink and one in green ink. The instant that Veronica opens her letter, she knows ....")

6.2 *Quantal states for distant measurements.* Mr. Parker is an intelligent layman. He is interested in quantum mechanics and is open to new ideas, but he wants evidence before he will accept wild-eyed assertions. "I like the argument of experiment 6.1," he says, "but I don't like the idea that when the left atom is detected, the right atom instantly jumps into the state with $m_z = -m_B$. I think that one atom is produced in the state $m_x = +m_B$ and the other atom is produced in the state $m_x = -m_B$, and that there are no instant state jumps." Show that Mr. Parker's suggestion is consistent with the

---

¶ Perhaps by magic powers, but not so magic as to change the fact that the two atoms always leave from opposite exits.

observation that "the right atom leaves the + exit with probability $\frac{1}{2}$, and similarly for the left atom". However, show also that if it were true, then on about $\frac{1}{4}$ of the experimental runs, both atoms would emerge from their respective + exits.

6.3   *A probability found through quantum mechanics.* In the test of Bell's theorem, experiment 6.2, what is the probability given by quantum mechanics that, if the orientation settings are different, the two detectors will flash different colors?

6.4   *A probability found through local determinism.* The experimental test of Bell's theorem shows that the postulated instruction sets do not exist. But suppose that they did. Suppose further that a given source produces the various possible instruction sets with the probabilities listed below:

| kind of atom going left | probability of making such a pair |
|:---:|:---:|
| (RRR) | 1/2 |
| (RRG) | 1/4 |
| (GRR) | 1/8 |
| (RGG) | 1/8 |

If this particular source were used in experiment 6.2, what would be the probability that the detectors flash different colors? Hint: Compare the draft lottery problem 5.6.

# 7

# Variations on a Theme by Einstein

The previous chapter covered the most important aspects of the Einstein–Podolsky–Rosen conundrum. But some interesting new features have come up since Aspect performed his experiments, and I thought you might enjoy them, so I'll mention two of them here. You may skip this chapter without interrupting the flow of the book's argument.

The results of the Aspect experiment were welcomed by most scientists as a final confirmation of the principles of quantum mechanics, principles that had already been verified magnificently in numerous experiments that were not as clean nor as easy to understand as the test of Bell's theorem. But scientists also looked for possible flaws in the confirmation, and they found one. We have discussed an ideal experiment, in which the source produces a pair of atoms and each tilting analyzer detects one of them. But in Aspect's real experiment, it often happened that after the source launched its atoms only one of the two atoms was detected, and sometimes neither of them were. This is not surprising: perhaps one of the atoms collided with a stray nitrogen molecule and was deflected away from its detector, or perhaps the detector electronics were pausing to reset after detecting one atom when a second atom rushed in. For these reasons, in analyzing his experiment Aspect ignored cases where only one atom was detected. But another possibility is that each atom is generated with an instruction set which could include the instruction "don't detect me". If this possibility is admitted, then one can invent local deterministic schemes that are consistent with Aspect's experimental results.

Personally, I regard this objection as far-fetched. But either of the two proposed experiments described here would overrule this objection definitively, because both of them produce situations in which quantum mechanics predicts that something might happen, whereas local determinism predicts that the same thing will *never* happen. Neither experiment has been executed in its entirety, but work is in progress on both and the preliminary results announced to date support quantum mechanics and oppose local determinism.

## 7.1   The Greenberger–Horne–Zeilinger variation on the Einstein–Podolsky–Rosen experiment

This experiment involves a source that ejects three atoms in an initial state that is hard to produce and even harder to describe. It is impossible for me to justify the prediction of quantum mechanics in a book at this level. For these two reasons I considered ignoring this experiment altogether in writing this book. But there is a payoff so rich that I had to include it: Whereas the test of Bell's theorem gives a circumstance in which the quantal probability for something happening is 50% while the local deterministic probability is more than 55%, the Greenberger–Horne–Zeilinger (or GHZ) variation gives a circumstance in which the quantal probability is 1 and the local deterministic probability is 0.

A top view of the Greenberger–Horne–Zeilinger experiment is sketched below. The source ejects three atoms in a special state, and each atom flies

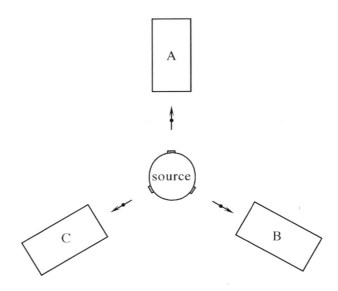

off to its own detector. Like the detectors in the test of Bell's theorem, each box contains a Stern–Gerlach analyzer that can be tilted and set to various orientations. But unlike the tilting analyzers used before, these analyzers can be set to only two orientations: the $z$ direction (vertical) or the $x$ direction (horizontal).

Back panel of each Greenberger–Horne–Zeilinger detector, showing the two orientations for its internal Stern–Gerlach analyzer. This analyzer is set to orientation $z$.

The orientations of the three analyzers are reported through a code like $xxz$, which means that detectors A and B are set to $x$ while detector C is set to $z$. As with the test of Bell's theorem (experiment 6.2, page 42), the detector orientations can be set after the atoms have been launched, while they are still in flight toward the detectors.

The predictions of quantum mechanics are:

|     | detector settings | what happens |
| --- | --- | --- |
| (1) | $zxx$ | odd number (1 or 3) go to + |
| (2) | $xxz$ | odd number (1 or 3) go to + |
| (3) | $xzx$ | odd number (1 or 3) go to + |
| (4) | $zzz$ | even number (0 or 2) go to + |
| (5) | other | not used in this argument |

Thus whenever two analyzers are set to $x$ and one to $z$, either all three atoms leave through the + exits of their respective analyzers, or else one leaves through the + exit and the other two leave through − exits.

### The argument for instruction sets

I will give an argument based on line (1) of the prediction that makes it seem reasonable that each atom is launched from the source with an instruction set, so that it will know whether to go to + or to − when it reaches its detector, regardless of what the settings of the detectors are. If you find this assertion reasonable already, you may skip the argument. Remember, however, that quantum mechanics maintains that this natural surmise is *not* correct, because an atom with a definite value of $m_x$ does not have a definite value of $m_z$.

Suppose that I wished to measure the value of $m_x$ for the atom going to detector C. One way to do it would be by setting A to $z$, B to $x$, and C to $x$, corresponding to line (1) of the prediction. Then I ask what would happen if I used only the detectors at A and at B, and forgot about the detector at C. (This despite the fact that it is the atom going to C that I'm interested in.) If detector A (set to $z$) measured +, and detector B (set to x) measured +, then detector C (set to x) would have to measure + as well, because according to line (1) of the prediction there must be either one or three atoms going to +. So if the atoms going to A and B come out through the + exit, then I don't need to actually measure $m_x$ of the atom going to C — I know what's going to happen at C merely by observing what had happened at A and B.

In fact, the same is true regardless of how the atoms come out at A and B, as long as the detectors are set to *zxx*:

|  | outcomes at A | B |  | outcome at C |
|---|---|---|---|---|
|  | + | + |  | + |
| given | + | − | then | − |
|  | − | + |  | − |
|  | − | − |  | + |

In short, if the settings are *zxx*, then by reading the outcomes at A and B, I can determine the outcome at C. I don't need to actually put an analyzer at C. The same is true for other directions: reading the outcomes at B and C enables me to determine the outcome at A, and reading the outcomes at A and C enables me to determine the outcome at B. And a glance at the quantal prediction on page 51 will convince you that parallel statements hold if the settings are *xxz* or *xzx*. In short, lines (1), (2), and (3) of the prediction enable you to determine either $m_z$ or $m_x$ of any atom merely by measuring appropriate quantities for the other two atoms, without actually touching the atom in question.

Because the detectors don't communicate with each other, the natural interpretation of this fact is that when an atom is launched from the source, it must already "know" how it will behave at the detector, regardless of the setting of that detector. Such an "instruction set" might be encoded into the direction of the atom's magnetic arrow, but it could conceivably be encoded in some strange or complicated way. In what follows I make no assumption about how the instruction set is encoded, only that it exists.

### *The prediction of local determinism*

I will write down the instruction set of all three atoms using a symbol like

atom heading toward
A  B  C
$$\begin{pmatrix} + & - & - \\ - & - & + \end{pmatrix} \quad \begin{matrix} \leftarrow \text{ if set to } z \\ \leftarrow \text{ if set to } x \end{matrix}$$

This notation means that the atom heading toward detector A will leave through the + exit if that detector is set to z, through the − exit if it is set to $x$. The atom heading toward detector B will leave the $-$ exit regardless of setting. The atom heading toward detector C will leave the $-$ exit if that detector is set to $z$, the $+$ exit if it is set to $x$.

Now I ask: What instruction sets are consistent with the quantal prediction? We will examine the first four lines of the table on page 51 in turn.

Line (1) of the table pertains to detector settings $zxx$, so it has nothing to say about what will happen if A is set to $x$, if B is set to $z$, or if C is set to $z$. In the following table the instructions for such settings are set to "?". Notice from the table that with these settings either one or three atoms leave through the $+$ exit, and therefore the only instruction sets compatible with line (1) are the following:

| instruction sets consistent with line (1) |
|:---:|

$$\begin{pmatrix} + & ? & ? \\ ? & - & - \end{pmatrix} \quad \begin{pmatrix} - & ? & ? \\ ? & + & - \end{pmatrix}$$

$$\begin{pmatrix} - & ? & ? \\ ? & - & + \end{pmatrix} \quad \begin{pmatrix} + & ? & ? \\ ? & + & + \end{pmatrix}$$

Which of these instruction sets is consistent with line (2) of the quantal prediction as well? We begin by considering only instruction sets of the type shown in the upper left above. Line (2) involves the setting $xxz$, so this reasoning will enable us to fill in the $x$ (bottom) slot of column A and the $z$ (top) slot of column C. We already know, from the entry above, that the atom heading for detector B will come out through the $-$ exit. Since a total of either one or three atoms must come out through the $+$ exit in this circumstance, then of the atoms heading for A and C, one must come out through $+$ and the other through $-$. Thus the instruction set must be either

$$\begin{pmatrix} + & ? & + \\ - & - & - \end{pmatrix} \quad \text{or} \quad \begin{pmatrix} + & ? & - \\ + & - & - \end{pmatrix}.$$

The same game can be played with the other three types of instruction sets consistent with line (1), resulting in:

| instruction sets consistent with lines (1) and (2) |
|:---:|

$$\begin{pmatrix} + & ? & + \\ - & - & - \end{pmatrix} \begin{pmatrix} + & ? & - \\ + & - & - \end{pmatrix} \begin{pmatrix} - & ? & - \\ - & + & - \end{pmatrix} \begin{pmatrix} - & ? & + \\ + & + & - \end{pmatrix}$$

$$\begin{pmatrix} - & ? & - \\ + & - & + \end{pmatrix} \begin{pmatrix} - & ? & + \\ - & - & + \end{pmatrix} \begin{pmatrix} + & ? & - \\ - & + & + \end{pmatrix} \begin{pmatrix} + & ? & + \\ + & + & + \end{pmatrix}$$

From here it is easy to find the instruction sets consistent with line (3) of the quantal prediction as well:

| instruction sets consistent with lines (1), (2), and (3) |
|---|

$$\begin{pmatrix} + & + & + \\ - & - & - \end{pmatrix} \quad \begin{pmatrix} + & - & - \\ + & - & - \end{pmatrix} \quad \begin{pmatrix} - & + & - \\ - & + & - \end{pmatrix} \quad \begin{pmatrix} - & - & + \\ + & + & - \end{pmatrix}$$

$$\begin{pmatrix} - & + & - \\ + & - & + \end{pmatrix} \quad \begin{pmatrix} - & - & + \\ - & - & + \end{pmatrix} \quad \begin{pmatrix} + & - & - \\ - & + & + \end{pmatrix} \quad \begin{pmatrix} + & + & + \\ + & + & + \end{pmatrix}$$

These eight, now completely determined, instruction sets are the only ones consistent with the quantal predictions given in lines (1), (2), and (3).

Which of these eight instruction sets is consistent with line (4) as well? In line (4) the detectors are set to *zzz*, so only the upper row of the instruction sets are relevant. The instruction set shown in the upper left above would result in all three atoms leaving through the + exits of their analyzers. But according to quantum mechanics (see line (4) of the quantal prediction on page 51), in this case an even number of atoms must leave + exits. Three is an odd number, so the instruction set in the upper left above must be ruled out as inconsistent with the predictions of quantum mechanics. The instruction set in the lower right must be ruled out for the same reason. All the remaining instruction sets call for exactly one of the three atoms to leave through + exits. But one is also an odd number! In short:

| instruction sets consistent with lines (1), (2), (3), and (4) |
|---|

# NONE!

Once again, the existence of instructions sets — regardless of how subtly the instructions are encoded — is inconsistent with the predictions of quantum mechanics.

## 7.2   Hardy's variation on the Einstein–Podolsky–Rosen experiment

This variation is harder to describe and I will not treat it in detail. It involves a source that ejects two atoms toward two different detectors, each of which can be tilted to two different angles, and an unusual initial state at the source. The experiment looks for a certain combination of events. The local deterministic prediction is that this combination will never happen. The quantal prediction is that it will happen with a

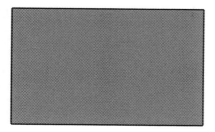

Fig. 7.1.   The golden rectangle.

probability of 9.017%. Thus if the combination happens in an experiment even once, then local determinism must be wrong.

One thing that intrigues me about this variation is the mathematical origin of the probability 0.09017.... The number is $g^5$, where the constant $g$ is equal to $(\sqrt{5} - 1)/2 = 0.6180\ldots$ and is called "the golden mean". If a line of length 1 is divided into two pieces so that the ratio of the length of the whole to the length of the long piece is equal to the ratio of the length of the long piece to the length of the short piece, then the long piece will have length $g$. The ancient Greeks considered a rectangle of width 1 and height $g$ to be the "ideal" (most beautiful) rectangle. The Parthenon in Athens, for example, has a height of $g$ times its width. Rectangles with these proportions also appear in the work of Leonardo da Vinci, Titian, and Mondrian. In addition the number is connected with the Great Pyramid, the star pentagram (which in one form appears in the American flag and which in another is said to call up the devil), the Fibonacci sequence, recursion relations, and with algorithms for locating the minimum of a one-variable function. But this is the first time I've ever seen it appear in quantum mechanics.

## 7.3   References

For the Greenberger–Horne–Zeilinger variation, see

Daniel M. Greenberger, Michael A. Horne, Abner Shimony, and Anton Zeilinger, "Bell's theorem without inequalities", *American Journal of Physics*, **58** (1990) 1131–1143,

N.D. Mermin, "The (non)world (non)view of quantum mechanics", *New Literary History*, **23** (1992) 855–875,

D. Bouwmeester, J.-W. Pan, M. Daniell, H. Weinfurter, and A. Zeilinger, "Observation of three-photon Greenberger–Horne–Zeilinger entanglement", *Physical Review Letters*, **82** (1999) 1345–1349.

For Hardy's variation, see

Lucien Hardy, "Nonlocality for two particles without inequalities for almost all entangled states", *Physical Review Letters*, **71** (1993) 1665–1668,

David Branning, "Does nature violate local realism?", *American Scientist*, **85** (1997) 160–167.

A technical but insightful exchange concerning Hardy's variation and its implications for locality in quantum mechanics is

Henry P. Stapp, "Nonlocal character of quantum theory", *American Journal of Physics*, **65** (1997) 300–304,

N. David Mermin, "Nonlocal character of quantum theory?", *American Journal of Physics*, **66** (1998) 920–924,

Henry P. Stapp, "Meaning of counterfactual statements in quantum physics", *American Journal of Physics*, **66** (1998) 924–926.

# 8

# Optical Interference

## 8.1 Overview

We have uncovered the first central principle of quantum mechanics, which is that *the outcome of an experiment cannot, in general, be predicted exactly; only the probabilities of the various outcomes can be found.* In particular, for the magnetic arrow of a silver atom, we know:

> If $m_z$ has a definite value, then $m_x$ doesn't have a value. If you measure $m_x$, then of course you find some value, but no one (not even the atom itself!) can say with certainty what that value will be — only the probabilities of measuring the various values can be calculated.

How do you like it? Do you feel liberated from the shackles of classical determinism? Or do you feel like Matthew Arnold, who wrote in *Dover Beach* that

> ... the world, which seems
> To lie before us like a land of dreams,
> So various, so beautiful, so new,
> Hath really neither joy, nor love, nor light,
> Nor certitude, nor peace, nor help from pain;
> And we are here as on a darkling plain
> Swept with confused alarms of struggle and flight,
> Where ignorant armies clash by night.

Regardless of your personal reaction, it is our job as scientists to *describe* nature, not to dictate to it!

In particular, we know that the model of a magnetic needle as an arrow, so carefully developed in chapter 2 and so correct within the domain of classical mechanics, must be wrong. In classical mechanics, magnetic needles behave like pointy sticks that precess in uniform magnetic fields and that both precess and move in non-uniform magnetic fields. We know that they don't behave this way in quantum mechanics, but we don't yet know how they *do* behave. We begin our search for their true behavior by examining what will turn out to be an analogous phenomenon, namely interference in light.

## 8.2    The interference of light

Light does not always travel in straight lines. You can demonstrate this for yourself with no more equipment than your hand and a street lamp. Go out on a dark night and look at the street lamp through a V formed by two of your fingers. Bring your fingers closer together to close the gap of the V. Just before your fingers touch and totally block your view of the street lamp, you will see the image of the street lamp become wider and wider as the gap between your fingers becomes narrower and narrower. (Alternatively, squint at the street lamp — as your eye lids grow very close together, the image of the street lamp grows very broad.) This is because light "spreads out" when it passes through a very narrow slit:

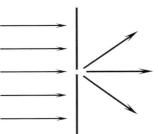

Even more remarkable is what happens when light passes through *two* adjacent narrow slits, a phenomenon called "two-slit interference":

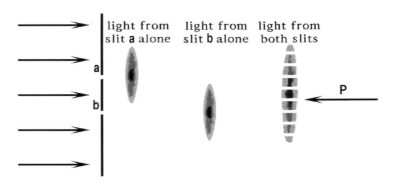

If only slit a were open, the light would spread out, as we have just seen, to make a wide bright band centered behind slit a; similarly for slit b. But if both slits were open, then the light would break up into a number of narrow, very bright bands separated by complete darkness. Notice particularly the situation at point P: This point is bright if either slit a or slit b is open, but dark if both slits are open! The term "interference" is quite appropriate for the phenomenon at this point: the light coming from slit b does not "cooperate" with the light from slit a to make brightness at point P, instead it "interferes" with the light provided by a to produce darkness.

You can demonstrate interference at home also, although the demonstration applies to light passing through many slits rather than through just two slits. (The results for the two different cases are actually quite similar.) A feather contains many parallel narrow slits. If you view a street lamp through a feather, you will see several images of the street lamp located side by side, and separated by darkness at points like point P.

Any explanation/description/recipe for this phenomenon must allow two sources of light to add up to darkness. I will describe two possibilities: the imaginary undulation and the imaginary stopwatch hand.

## 8.3    The undulation picture

In this picture each slit acts as a source of imaginary undulations —— like water waves except that there's no water. The undulations are close together: 15 800 wave crests per centimeter for red light, somewhat more for other colors. The total "water surface motion" is the sum of the undulations from each source. The sensation of light brightness at a point is due not to the height of the "water" there, but due to the difference in height from crest to trough there. (Quantitatively, in fact, the brightness is proportional to the square of that difference.)

The figure below is a schematic diagram of the imaginary undulations

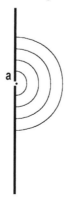

produced by a single slit at a given instant. The thin lines mark the wave crests, the troughs (not shown in the diagram) fall half-way between the crest lines. As time goes on the wave crests travel to the right, and new crests emerge from the slit source to replace them. (At the particular instant shown in this snapshot, a new crest is just about to emerge from the slit.)

The figure below similarly shows the imaginary undulations produced by a different single slit, located somewhat lower.

What happens when both slits are present? The figure below is similar to the two above in that it shows the situation at a single instant, but it differs in that the circles radiating from the two slits do not mark wave crests. Instead the circles radiating from slit a mark where the wave crests would be if only slit a were present, and those radiating from slit b mark where they would be if only slit b were present. The actual status of the water surface, i.e. the total "water surface motion", must be found by summing the undulation from slit a and the undulation from slit b.

Point Q is located exactly half-way between slit a and slit b, so when a crest from a arrives there a crest from b arrives also, and the "water level"

is very high. Similarly the troughs from a reach point Q at the same time that the troughs from b do, so then the "water level" is very low. Thus the "water surface" rises and falls dramatically at point Q, corresponding to intense brightness there.

The situation at point P is very different. It is somewhat closer to slit a than it is to slit b, so when a crest from slit a arrives, the corresponding crest from slit b is still in transit, instead the contribution from slit b is the preceding trough! At point P the crests from a arrive on top of the troughs from b, and the troughs from a arrive on top of the crests from b. Indeed, the contributions from the two slits exactly cancel out at all times, so the "water surface" does not move at all, corresponding to complete darkness. Now you can see how, in this picture, two sources of light can interfere to produce darkness.

## 8.4 The stopwatch hand picture

In this picture each slit sends out streams of "photons" ("particles of light"). When the slit releases a photon, an imaginary stopwatch hand starts moving. For red light, the hand rotates 15 800 times every time the photon moves one centimeter. To find the brightness at any point, add the two stopwatch hands (one from each slit) by laying them tail to head. The "sum" stretches from the tail of the first stopwatch hand to the head of the second stopwatch hand. The brightness at that point equals the square of the magnitude (i.e. the "length") of the sum.

It is traditional to start each stopwatch hand pointing to the right, i.e. to 3 o'clock, and to rotate it counterclockwise, but this is only convention. Any other convention, as long as it is applied consistently, will find the same resulting brightness pattern. A stopwatch hand is also called a "rotating arrow", a "phasor", or, by the cognoscenti, a "complex number".

For example, in the two-slit situation described above, the stopwatch hand associated with photons that travel from a to Q starts pointing at 3 o'clock, rotates three times, and ends up pointing to 3 o'clock again. (This is because the distance from a to Q is exactly 3/(15 800) centimeters.) Similarly for the stopwatch hand associated with photons that travel from

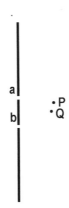

b to Q (which is exactly the same distance). The sum arrow is a long one, corresponding to intense brightness.

Turning our attention now to point P, we find that the arrow associated with photons traveling from a to P rotates three complete times, stopping at 3 o'clock. (The distance from a to P is again 3/(15 800) centimeters.) Meanwhile the arrow associated with photons traveling from b to P rotates three and a half times, stopping at 9 o'clock. (Because the distance from b to P is 3.5/(15 800) centimeters.) These two arrows add to zero,

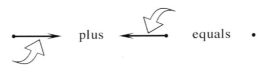

corresponding to complete darkness. Again you see how this picture, like the wave picture, permits two sources of light to interfere and produce darkness.

## 8.5   Philosophical remark

There are no stopwatch hands, just as there is no water. Both of these pictures are nothing but analogies — mathematical schemes that permit us to calculate the brightness of the light striking various points. Yet both pictures give complete and accurate descriptions of the behavior of light. One scheme cannot be preferred over the other on scientific grounds, because both give exactly the same results for the brightness. Neither scheme gives an underlying mechanism* that tells us "what's really going on". Neither, I suppose, is what God was thinking when he/she/it created the universe. If you want the answers to such questions, you must consult a priest, not a scientist.

## 8.6    Problems

8.1   *Adding arrows.* Three stopwatch hour hands each have a length of five inches. One stopwatch hand points to noon ("due north"), the second to 3 o'clock ("due east"), and the third to 1:30 ("northeast"). How long is the sum of the three arrows, and in which direction does it point? Hint: This is the $1:1:\sqrt{2}$ right triangle again.

8.2   *Philosophical remark.* Here are three different ways to add seven and sixteen: (1) Use arabic numerals 7 and 16. (2) Use roman numerals VII and XVI. (3) Put seven marbles in a box, put in sixteen more, then count all the marbles in the box. Which process is "really going on" in addition?

---

* For an insightful discussion about mathematical algorithms *vs.* clockwork mechanisms, see chapter 2, "The relation of mathematics to physics" of R.P. Feynman, *The Character of Physical Law* (MIT Press, Cambridge, Massachusetts, 1965).

# 9

# Quantal Interference

We have seen that quantum mechanics can only find probabilities and not certainties. Now we must find out how to work with these probabilities.* We will do this by examining the results of several experiments performed with a new instrument, the *interferometer* (also called an *analyzer loop*).

The interferometer is a Stern–Gerlach analyzer followed by plumbing that recombines the paths of atoms leaving from either exit. The design

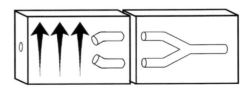

above is represented by the simple figure below. An interferometer must be

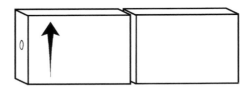

constructed in such a way that the two branches are absolutely identical, whence it is impossible to tell by examining the outgoing atom which of the two branches it went through. For example, the two branches must have exactly the same length, because otherwise it would take an atom more time to traverse the longer branch. Because of this precise construction,

---

* This book presents the standard description of quantum mechanics. Other descriptions — notably that of David Bohm — are also possible. But, as required by the Einstein–Podolsky–Rosen effect, all of the viable alternative descriptions are either probabilistic or non-local or both.

when an atom leaves the interferometer it is in exactly the same state as it was when it entered. This holds regardless of the interferometer's orientation.

Thus the interferometer is an instrument that does nothing at all! The outgoing atom is the same as the incoming atom. It is hard to see why anyone would want to build one. Of course it can be made to do something useful by blocking one of its two branches. For example, in the interferometer below the lower branch is blocked, so it behaves just like a vertical Stern–Gerlach analyzer with its bottom exit blocked: not all of the incoming atoms will go out, but each one that does has $m_z = +m_B$.

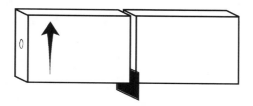

I will describe several experiments using the apparatus sketched below. In all cases the input atom has $m_z = +m_B$ (it has been gathered from the + exit of a vertical analyzer not shown in the figure). The atom passes through a horizontal interferometer, and then it is analyzed with a vertical analyzer. An atom leaving the − exit of the vertical analyzer is considered output, while an atom leaving the + exit is ignored.

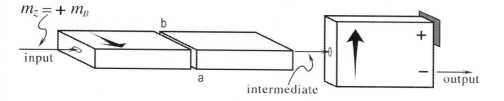

## 9.1 Experiment 9.1: Branch a is blocked

If branch a is blocked, then:

> The probability of passing from input to intermediate is $\frac{1}{2}$.
> The intermediate atom has $m_x = -m_B$.
> The probability of passing from intermediate to output is $\frac{1}{2}$.
> The overall probability of passing from input to output is $\frac{1}{2} \times \frac{1}{2} = \frac{1}{4}$.

## 9.2    Experiment 9.2: Branch b is blocked

If branch b is blocked, then the experiment proceeds exactly the same as experiment 9.1, except that the intermediate atom has $m_x = +m_B$.

## 9.3    Experiment 9.3: Neither branch is blocked

*Analysis A.* (Using the laws for compound probability.)
>      The atom goes from input to output either through branch a or through branch b.
>      It goes through branch a with probability $\frac{1}{4}$, or through branch b with probability $\frac{1}{4}$, so the overall probability of passing from input to output is $\frac{1}{4} + \frac{1}{4} = \frac{1}{2}$.

*Analysis B.* (Using the fact that an interferometer passes atoms unchanged.)
>      The probability of passing from input to intermediate is 1.
>      The intermediate atom has $m_z = +m_B$.
>      Any such atom leaves the + exit of the vertical analyzer, so ...
>      The overall probability of passing from input to output is 0.

A monumental disagreement! Which analysis is correct? Experiment confirms the result of analysis B, but what could possibly be wrong with analysis A? Certainly $\frac{1}{4} + \frac{1}{4} = \frac{1}{2}$ is correct, certainly the rule for compound probability (which is embodied in the second sentence) is correct. The only possible error is in the first sentence: "The atom goes either through branch a or through branch b." This common-sense assertion must be wrong! Indeed, if the atom passed through branch a then at the intermediate stage it would have a definite value of $m_x = +m_B$, but we know that this intermediate atom has a definite value of $m_z$ so it *can't* have a definite value of $m_x$. The interferometer, which seemed so useless just a moment ago, is in fact an extremely clever way of correlating the position of an atom with its $m_x$: if $m_x = +m_B$, then the position is in branch a; if $m_x = -m_B$, then the position is in branch b. Since the incoming atom lacks a definite value of $m_x$, it must lack a definite position as well. The English language was invented by people who did not understand quantum mechanics, so it doesn't have an accurate concise way to describe what is going on in this experiment. The best approximate phrase is "the atom goes through both branches".

This conclusion seems patently absurd. Actually it is correct, and it seems absurd only if one thinks of an atom as being like a marble, only infinitely smaller and infinitely harder. In fact an atom is no more a small hard marble than an atom's magnetic needle is a pointy stick. These classical ideas are simply wrong when applied to very small objects. But I don't expect you to take my word for it. Let's perform an experiment in which we actually look at the two branches to see whether the atom is going through branch a, branch b, or both branches.

## 9.4 Experiment 9.4: Watching for atoms

In this experiment neither branch is blocked, but we train a powerful lamp on each branch to see whether the atom passes through branch a or through branch b. Inject an atom into the apparatus — a moment later we see a glint of light at branch b: the atom is going through branch b. Another atom, a glint at b again. Then a glint at a, then b again, then at a, etc. Never do we see, say, two weak glints, one at a and the other at b. "Ah ha!" you say, "So much for your metaphysical nonsense, Mr. Styer. Our observations show that the atom is going either through branch a or through branch b, and never through 'both', whatever *that* may mean."

True. But now look at the probability of passing from input to output. For unwatched atoms (experiment 9.3), that probability is zero. For watched atoms (experiment 9.4), that probability is $\frac{1}{2}$. If an atom *is* watched, then it *does* go either through branch a or through branch b, analysis A *is* correct, and half the atoms *do* leave the output! In fact, when the glint is seen at branch a then the intermediate atom has $m_x = +m_B$, as can be confirmed by replacing the vertical Stern–Gerlach analyzer with a horizontal one: an atom that causes a glint at branch a will always leave through the + exit of a horizontal analyzer, while one that causes a glint at branch b will always leave through the − exit.

Clearly a "watched" (or "observed") atom behaves differently from an unwatched atom. Much silliness has been written concerning the subject of precisely what constitutes an observation. Suppose, for example, that we train the lamps on the interferometer but turn our backs and don't look for the glints. Have the atoms been watched or haven't they? What if the glints are watched by cats rather than by human beings? Such questions are most easily answered by considering a parallel experiment. Suppose we turn our backs on the glints but record them on a movie. Now suppose the movie is played back, to either a human or a feline audience, one hour after the experiment is finished. Certainly by this time it is too late to change the way atoms exit from the vertical analyzer! In fact the significant question is not whether someone actually *sees* which branch an atom takes, but whether it is, in principle, *possible* to determine which branch an atom takes, regardless of whether any human actually takes advantage of that possibility. (Sometimes the term "registered" is used instead of "observed" or "measured" to emphasize that no human involvement is required.) From this perspective, the blocks in experiments 9.1 and 9.2 are simply ways to determine which branch the atom took: if the atom emerges while branch a is blocked, then it must have taken branch b. (I warn you, however, that it is not always easy to decide whether or not an observation is "in principle possible", nor to uncover the exact moment at which an observation is made.)

Perhaps you think that the "problem" with experiment 9.4 is that the atoms are being disturbed by the intense light. An atom is a tiny thing, after all, and perhaps the blast of light is simply pushing it around uncontrollably. This thought inspires the next two experiments.

### 9.5    Experiment 9.5: Watching for atoms at branch a only

In this case the intense light is trained only on branch a, so it cannot possibly disturb an atom that passes through branch b. As an atom passes through the interferometer there is either a glint at a, which means that the atom has passed through branch a, or else there is no glint at all, which means that the atom has passed through branch b. Since it is possible to determine which branch the atom passed through, the results are exactly the same as those of experiment 9.4.

### 9.6    Experiment 9.6: Watching for atoms with dim light

Although the light is dimmer, the glints are exactly the same! (This is because each glint corresponds to exactly one photon.) When the light is dim, however, some atoms pass through the interferometer without producing a glint at all. Careful analysis of the experimental results shows that an atom which produces no glint behaves just as if it were in experiment 9.3 (unwatched atoms), while one which does produce a glint behaves just as if it were in experiment 9.4 (watched atoms).

### 9.7    Is measurement magical?

How can the behavior of an atom depend upon whether or not it is being watched? Can't watching happen without the atom being affected? No. The only way to observe/measure/watch a system is to influence/disturb/alter it in some way. Consider, for example, a ball tossed upward in a room with ceiling lamps. If the lamps are off, the ball will ascend to a certain height. If the lamps are on, then the light will press down on the ball and it will attain a somewhat lower height. This effect is negligible if the ball is a baseball[†] but important if the ball is an atom, because it is much easier to push an atom around than a baseball. (Notice that it is the presence of light, not of watchers, in the room that makes the difference. Once again, the important issue is whether the observation is possible in principle, not whether a person — or a cat — happens to take advantage of that possibility.)

This is not to say that all questions concerning quantal measurement — and concerning its sister subject, the classical limit of quantum mechanics — are completely solved and pat. They are not. Consider the question of the Stern–Gerlach analyzer *vs.* the Stern–Gerlach interferometer. In the first device, the atom emerges from one exit or the other but not both. In the second device, the atom goes through one branch or the other or both. But the front half of an interferometer is exactly the same as an analyzer! How does the atom "know" that in the interferometer the two branches will ultimately be recombined?[‡] Questions like these are far more subtle than they appear, and are the subject of current investigation. Although measurement is not magical, it still holds mysteries.

## 9.8 Understanding

Whenever I lecture concerning the topic of this chapter, students approach me afterwards and say "I followed the lecture, but I just don't understand it." When I delve into exactly what is disturbing these students, it usually turns out to be one of two conceptual roadblocks: either the student simply finds that this behavior is unfamiliar and unexpected, or else (s)he is seeking a mechanism which underlies the behavior.[§]

This behavior certainly is unexpected, but that doesn't mean that it is wrong. If you were born in orbit in a space station and landed on earth for your sixteenth birthday, then you would find gravitational attraction unfamiliar and unexpected. But it is not wrong to feel that way. Indeed gravity truly is a mysterious force! Many people feel more comfortable with a new phenomenon if it is given a name. The strange attraction of remote bodies is called "gravity". Perhaps it will comfort you to know that the strange phenomenon described in this chapter is called "quantal interference".

---

[†] Indeed, the effect is small enough that many people don't know it exists. However, all science fiction buffs have read stories about spaceships driven by the sunlight reflected from huge gossamer sails.

[‡] This is the content of the so-called "Schrödinger's cat" paradox.

[§] Another discussion of the meaning of "understanding" in science is given by R.P. Feynman in *QED: The Strange Theory of Light and Matter* (Princeton University Press, Princeton, New Jersey, 1985), pages 9–10.

What is the mechanism that underlies quantal interference? People ask
this question thinking that there is some explanation of the sort: "An atom
is made up of two bricks held together with a rubber band, and when
the rubber band hits the wall of branch a then the two bricks oscillate
back and forth and ... ". But an atom is not made up of bricks and
rubber bands. Instead bricks and rubber bands are made up of atoms!
The Einstein–Podolsky–Rosen arguments show that no local deterministic
mechanism, no matter how intricate, can lead to the results of quantum
mechanics. As far as anyone knows, there is no mechanism. This is simply
the way the universe works.

### 9.9    References

The idea that interference lies at the heart of quantum mechanics was
recognized from the the founding of the subject in the 1920s, but it has
been emphasized most notably in theoretical treatments by Feynman. See,
for example,

> R.P. Feynman, *The Character of Physical Law* (MIT Press, Cam-
> bridge, Massachusetts, 1965) chapter 6,
>
> R.P. Feynman, *QED: The Strange Theory of Light and Matter* (Prince-
> ton University Press, Princeton, New Jersey, 1985) pages 77–82,
>
> R.P. Feynman, R.B. Leighton, and M. Sands, *The Feynman Lectures
> on Physics* volume III: *Quantum Mechanics* (Addison-Wesley,
> Reading, Massachusetts, 1965) chapters 1 and 5,
>
> R.P. Feynman and A.R. Hibbs, *Quantum Mechanics and Path Inte-
> grals* (McGraw-Hill, New York, 1965) chapter 1.

Interference experiments using photons have been performed in the laboratory for centuries. But laboratory (as opposed to "thought experiment")
interferometers that use matter rather than light are relatively young. An
accessible description of an early experiment is

> D.M. Greenberger and A.W. Overhauser, "The role of gravity in
> quantum theory", *Scientific American*, **242** (5) (May 1980) 66–76,
> 186.

This interferometer employed neutrons and its builders used it to investigate the effects of gravity. Interferometers using matter have grown
steadily more sophisticated. This growth is reviewed in

> Barbara Levy, "Atoms are the new wave in interferometers", *Physics
> Today*, **44** (7) (July 1991) 17–20,

and it has culminated, at least for the moment, in the actual execution of
the experiments suggested so long ago as theoretical exercises by Feynman:

A. Tonomura, J. Endo, T. Matsuda, T. Kawasaki, and H. Ezawa, "Demonstration of single-electron buildup of an interference pattern", *American Journal of Physics*, **57** (1989) 117–120,

R. Gähler and A. Zeilinger, "Wave-optical experiments with very cold neutrons", *American Journal of Physics*, **59** (1991) 316–324,

Michael S. Chapman, David E. Pritchard, *et al.*, "Photon scattering from atoms in an atom interferometer: Coherence lost and regained", *Physical Review Letters*, **75** (1995) 3783–3787,

E. Buks, R. Schuster, M. Heiblum, D. Mahalu, and V. Umansky, "Dephasing in electron interference by a 'which-path' detector", *Nature*, **391** (1998) 871–874.

The research questions concerning measurement and the classical limit, touched upon in section 9.7, are discussed in more detail and at various technical levels in

J.A. Wheeler and W.H. Zurek, editors, *Quantum Theory and Measurement* (Princeton University Press, Princeton, New Jersey, 1983) especially pages 184–185,

A.J. Leggett, "Schrödinger's cat and her laboratory cousins", *Contemporary Physics*, **25** (1984) 583–598,

Eric J. Heller and Steven Tomsovic, "Postmodern quantum mechanics", *Physics Today*, **46** (7) (July 1993) 38–46,

V.B. Braginsky and F.Ya. Khalili, "Quantum nondemolition measurements: the route from toys to tools", *Reviews of Modern Physics*, **68** (1996) 1–11,

Paul Kwiat, Harold Weinfurter, and Anton Zeilinger, "Quantum seeing in the dark", *Scientific American*, **275** (5) (November 1996) 72–78,

Serge Haroche, "Entanglement, decoherence and the quantum/classical boundary", *Physics Today*, **51** (7) (July 1998) 36–42.

Anyone who remembers the American presidential election of 1992 (Bush *vs.* Clinton *vs.* Perot) will enjoy the many insights, concerning both physics and politics, to be found in

N.D. Mermin, "Two lectures on the wave-particle duality", *Physics Today*, **46** (1) (January 1993) 9–11.

## 9.10  Sample problem

In the apparatus sketched on the next page, atoms with $m_z = +m_B$ are passed through a horizontal interferometer (number 1) then a vertical interferometer (number 2). If all branches are open, 100% of the incoming

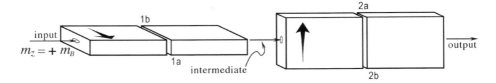

Fig. 9.1. Two interferometers. (Sample problem on page 71.)

atoms exit from the output. What percentage of the incoming atoms leave from the output if the following branches are blocked? (The atoms are not observed as they pass through the interferometers.)

(a)  2a                    (d)  1b
(b)  2b                    (e)  1b and 2a
(c)  1a                    (f)  1a and 2b

*Solution*

Only two principles are needed to solve this problem: First, an atom leaving an unblocked interferometer leaves in the same state that it was in when it entered. Second, an atom leaving an interferometer that has one branch blocked leaves in the state specified by the branch through which it passed, regardless of what its entry state was. Use of these principles gives the solution on page 73. Notice that in changing from situation (a) to situation (e), you add blockage, yet you increase the output!

## 9.11   Problems

9.1   *Terminology.* Why are the phenomena described in this chapter better called "atom interference" rather than "the interference of atoms"?

9.2   *A different interference setup.* If the apparatus sketched on page 65 were changed so that atoms leaving the − exit were ignored, and atoms leaving the + exit were considered output, then what would be the probability of an atom passing from input to output if (a) branch a were blocked, (b) branch b were blocked, or (c) neither branch were blocked.

9.3   *Three interferometers.* Atoms with $m_z = +m_B$ pass through a horizontal interferometer, then a vertical interferometer, then a horizontal interferometer, as shown on page 74. What percentage of the incoming atoms leave from the output if the following branches are blocked? (The atoms are not observed as they pass through the interferometers.)

| branches blocked | input state | branch taken through # 1 | intermediate state | branch taken through # 2 | output state | probability of input → output |
|---|---|---|---|---|---|---|
| none | $m_z = +m_B$ | both | $m_z = +m_B$ | a | $m_z = +m_B$ | 100% |
| 2a | $m_z = +m_B$ | both | $m_z = +m_B$ | 100% stopped at a | none | 0% |
| 2b | $m_z = +m_B$ | both | $m_z = +m_B$ | a | $m_z = +m_B$ | 100% |
| 1a | $m_z = +m_B$ | 50% stopped at a  50% pass through b | $m_x = -m_B$ | both | $m_x = -m_B$ | 50% |
| 1b | $m_z = +m_B$ | 50% pass through a  50% stopped at b | $m_x = +m_B$ | both | $m_x = +m_B$ | 50% |
| 1b and 2a | $m_z = +m_B$ | 50% pass through a  50% stopped at b | $m_x = +m_B$ | 25% stopped at a  25% pass through b | $m_z = -m_B$ | 25% |
| 1a and 2b | $m_z = +m_B$ | 50% stopped at a  50% pass through b | $m_x = -m_B$ | 25% pass through a  25% stopped at b | $m_z = +m_B$ | 25% |

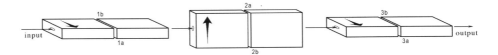

Fig. 9.2.   Three interferometers. (Problem 9.3.)

| (a) | 3a          | (d) | 2b         | (g) | 1b and 3b         |
|-----|-------------|-----|------------|-----|-------------------|
| (b) | 3b          | (e) | 1b         | (h) | 1b and 3a         |
| (c) | 2a          | (f) | 2a and 3b  | (i) | 1b and 3a and 2a  |

(Note that in going from situation (h) to situation (i) you get *more* output from *increased* blockage.)

9.4  *Paradox?*

   (a) The year is 1492, and you are discussing with a friend the radical idea that the earth is round. "This idea can't be correct," objects your friend, "because it contains a paradox. If it were true, then a traveler moving always due east would eventually arrive back at his starting point. Anyone can see that that's not possible!" Convince your friend that this paradox is not an internal inconsistency in the round-earth idea, but an inconsistency between the round-earth idea and the picture of the earth as a plane, a picture which your friend has internalized so thoroughly that he can't recognize it as an approximation rather than the absolute truth.

   (b) The year is 1992, and you are discussing with a friend the radical idea of quantal interference. "This idea can't be correct," objects your friend, "because it contains a paradox. If it were true, then an atom passing through branch a would have to know whether branch b were open or blocked. Anyone can see that that's not possible!" Convince your friend that this paradox is not an internal inconsistency in quantum mechanics, but an inconsistency between quantal ideas and the picture of an atom as a hard little marble that always has a definite position, a picture which your friend has internalized so thoroughly that he can't recognize it as an approximation rather than the absolute truth.

(If you cannot solve this problem now, then come back to it after reading section 15.2, "What does an electron look like?")

9.5  *Definite position.* "It is absurd," Mr. Parker says, "to think that an atom might not have a definite position. It's not just atoms and positions, but *anything* must have a definite value for *any* of its attributes." You know that a glass prism splits white light up into its component colors. Convince Mr. Parker that a prism doesn't have a definite color.

9.6 *Misconceptions.* In his book *In Search of Schrödinger's Cat*, John Gribbin describes an experiment similar to our interferometer experiments, and concludes that "unless someone looks, nature herself does not know which hole the electron is going through". Which two misconceptions are embodied in this sentence?

# 10
# Amplitudes

## 10.1    The amplitude framework

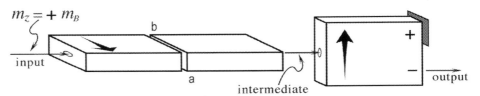

Recall from the first three quantal interference experiments (pages 65–67) that in the above apparatus, the probability of passing from the initial state (at input with $m_z = +m_B$) to the final state (at output with $m_z = -m_B$) is

| situation | probability |
|---|---|
| branch a open | 1/4 |
| branch b open | 1/4 |
| both branches open | 0 |

Clearly the probability of passing through both branches does not equal the sum of the probability of passing through branch a plus the probability of passing through branch b. On the other hand, it seems natural to ascribe the total probability to some sort of an "influence through branch a" plus an "influence through branch b". (Recall that optical interference was described by a similar picture, where the "influence through a slit" was either the undulation due to that slit or the stopwatch hand associated with a photon passing through that slit.) It somehow seems unscientific to call these things "influences", a word beloved by mediums and witches, so they are called "amplitudes" (or sometimes "probability amplitudes"). At the moment the existence of amplitudes is nothing but a reasonable surmise, but this guess will turn out to be an excellent one, supported by reams of evidence (to be reviewed later in this chapter). For now, however, our task is to firm up the concept of amplitudes, and, in particular, to find a mathematical representation for them.

The salient feature of amplitudes is that the sum of an "amplitude to pass through branch a" plus an "amplitude to pass through branch b" can lead to a total probability of zero. Thus an amplitude cannot be represented by an intrinsically positive number, because two positive numbers cannot add up to zero. There are, however, many classes of mathematical entities for which two elements of the class *can* add to zero. One such class is the real numbers, as demonstrated by $(+0.7)+(-0.7) = 0$. We will see in section 11.1 that the class of real numbers cannot adequately represent all possible amplitudes. Instead, amplitudes must be represented by two-dimensional arrows* similar to the rotating stopwatch hands of the optical interference experiment (section 8.4). If there are several ways of going from the initial to the final state, then the "total amplitude" for doing so is just the sum of the several individual amplitudes, where arrows are summed by placing them tail to head as described on page 61. The probability of going from the initial to the final state is just the square of the magnitude of the total amplitude arrow.

Let us see how the general ideas of amplitudes and probabilities presented above can explain the first three quantal interference experiments from the preceding chapter. The amplitude to go from input to output via branch a is represented by an arrow of magnitude $\frac{1}{2}$ pointing right: →. The amplitude to go from input to output via branch b is represented by an arrow of magnitude $\frac{1}{2}$ pointing left: ←. When both branches are open, the total amplitude is represented by the sum of the two arrows, which is just an arrow of magnitude zero.

| situation | sum of amplitudes | probability |
|---|---|---|
| branch a open | → | 1/4 |
| branch b open | ← | 1/4 |
| both branches open | · | 0 |

Now we can firm up the vague phrase "the atom goes through both branches" introduced in the last chapter. Its precise meaning is simply that there is an amplitude for the atom to go through either branch.

> *Technical aside:* The above paragraph illustrates important general techniques for assigning amplitude arrows. The magnitude of an arrow can be fixed by knowing the corresponding probability, because the magnitude is just the square root of the prob-

---

* These amplitude arrows are not related to the magnetic needle arrows introduced in chapter 2. This book represents magnetic needles by arrows with filled arrowheads and amplitudes by arrows with open arrowheads.

ability. (In the situation above, $\frac{1}{2} = \sqrt{\frac{1}{4}}$.) The angles between the arrows are harder to find: they must be uncovered through the results of interference experiments. Section 11.1 (page 86) works out such an assignment problem in some detail.

Amplitude arrows are mathematical tools that permit the computation of probabilities, they are not physical entities that are actually located in space and observable if only you were to look hard enough.[†] You must not think that there are two real live physical arrows out there, one flying through branch a and the other flying through branch b. For one thing, the amplitude arrows are dimensionless — an arrow is not $\frac{1}{2}$ inch long or $\frac{1}{2}$ millimeter long, it is just $\frac{1}{2}$ long. For another, the orientation of the arrows is not specified exactly. If each arrow in any given problem is rotated by the same angle, then the same probabilities will result. The association between the physical entity (an amplitude) and its mathematical representation (an arrow) is not unique.[‡] Finally, we will see in section 11.2 (page 91) that amplitude arrows must often be assigned to composite processes, such as the motion of two particles, where it is impossible to associate an amplitude with a single particle.

We have uncovered the second — and last — central concept of quantum mechanics: *The probabilities of various outcomes arise through the interference of amplitudes.* This is a good place to summarize our entire discussion.

---

### A summary of all quantum mechanics

The question of quantum mechanics:

What is the probability of going from one state to another?

The framework for answering that question:

(1) Enumerate all ways of going between the two states.
(2) Assign an amplitude (an arrow) to each way.
(3) Add up all the arrows (place arrows tail to head, the sum stretches from the first tail to the last head).
(4) The probability is the square of the magnitude of this sum arrow.

---

This list is a framework rather than an actual recipe for answering the question because it doesn't say how to perform the assignment of

---

[†] Indeed, it is possible to find schemes for calculating the outcome probabilities that do not make use of amplitude arrows at all. One such scheme — which is somewhat like the "water wave" scheme for calculating the interference effects of light — was invented by David Bohm.

[‡] This is not so unusual as you might at first think. For example, the relationship between lengths and positive numbers is not unique. The same length is represented by both 2 (feet) and 24 (inches).

amplitudes to ways required by point (2). Physics majors spend many years learning the rules for assigning amplitudes. (As well as learning how to guess which rules might apply in situations that have not yet been encountered!) For this book, I will just tell you the appropriate rules as they are needed. (If you ask your physicist friends about rules for assigning amplitudes, they won't know what you're talking about. That's because they use the technical phrase "the Hamiltonian (or the Lagrangian) for the system" instead of the phrase "the rules for assigning amplitude arrows".)

Another problem with implementing this framework is less obvious. What, precisely, is meant by a "state"? This is another question that can require considerable thought and experimentation to answer, and for which the answer is sometimes surprising. For the "unwatched" atoms considered so far in this chapter, the state is specified as, for example, "an atom leaving the − exit of the vertical analyzer". But for "watched" atoms, the state specification must give information about both the atom *and* the photon that interacted with it. This is how the results of quantal interference experiments 9.4 through 9.6 on pages 67–68 can be worked into this framework.

For example, part A of figure 10.1 (next page) shows an atom with $m_z = +m_B$ entering an interference apparatus while a photon approaches branch a to observe the atom. (In the figure, the atom is represented by a dot and the photon by a square.) If the atom is observed to pass through branch a (photon is deflected, as in situations B and C) then the intermediate atom has $m_x = +m_B$ and the atom could leave through either the + or the − exit of the vertical analyzer. If the atom is not observed (the photon misses, as in situation D) then the intermediate atom has $m_z = +m_B$ and the atom must leave through the + exit of the vertical analyzer. Thus there is some amplitude to go from state A to state B, and some amplitude to go from state A to state C, but no amplitude to go from state A to state D. But states B and D are exactly the same as far as the atom is concerned, they differ only in the photon. Thus to specify a "state" in this circumstance you must give the position of both the atom and the photon.

Finally, the framework is imprecise about the meaning of "way". Suppose an atom moves from point A to point B. This could be done through a direct, straight line route, or it could be done via a detour to London. Both of these paths are "ways" to perform the move and both must be considered. But there are other, less obvious, ways. For example, the atom could leave point A, move toward B, emit a photon, move toward B a little more, reabsorb that same photon, then continue its journey on to point B. Or it could leave point A intact, break into three pieces and then reassemble before getting to point B. Do such bizarre mechanisms constitute "ways to go from the initial to the final state"? Yes they do. Most of the time, however, such truly bizarre ways can be ignored for practical purposes

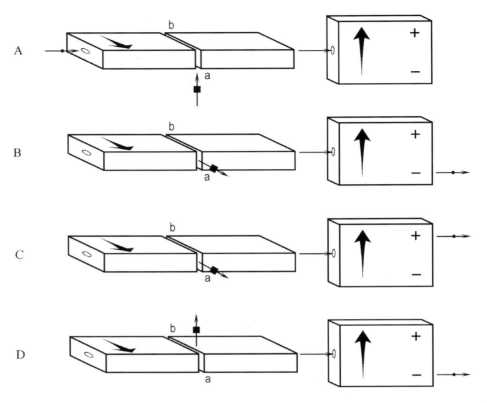

Fig. 10.1. Various states for an atom being observed as it passes through an interferometer. To specify a state, you must give the position of the photon (represented by a square) as well as the position of the atom (represented by a dot).

because (1) the arrows associated with such ways are quite small indeed and (2) there are a host of other ways that are similar to, say, the three-pieces way (for example, the atom breaks into four pieces) and the various arrows from this host of similar ways point in all different directions, so when they are all added together they tend to cancel each other out.

## 10.2   Evidence for the amplitude framework

In the Einstein–Podolsky–Rosen experiment we found a single definitive[§] experiment which proved that classical mechanics (or any other local

---

[§] That is, definitive except for the considerations mentioned on page 49. It is a characteristic of science that all experiments involve error and thus that no experiment — and no scientific statement — is *absolutely* definitive.

deterministic scheme) must be incorrect. It would be nice to present now a definitive experiment which proves that the amplitude framework is correct. *This cannot be done.* One experiment can prove a general idea wrong, but no number of experiments can prove that idea right. This is the nature of a general idea: it is supposed to work in all cases, so if it fails in a single test it must be wrong, but if it passes a million tests it might still fail the million and first test. (To prove that rhinoceroses exist, you only need to find one rhinoceros. To prove that unicorns do not exist, you need to scour the earth and find none.) Because general ideas cannot be proven correct, I will instead present an overview of the many and various situations to which the amplitude framework has been applied, and for which it has never yet been found wanting.

| object | approximate size |
|---|---|
| person | $10^0$ meter |
| fly | $10^{-2}$ meter |
| hair width | $10^{-4}$ meter |
| bacterium | $10^{-6}$ meter |
| DNA width | $10^{-8}$ meter |
| atom | $10^{-10}$ meter |
|  | $10^{-12}$ meter |
| nucleus | $10^{-14}$ meter |
|  | $10^{-16}$ meter |
|  | $10^{-18}$ meter |
| quark | $10^{-20}$ meter |
|  | $10^{-22}$ meter |
|  | $10^{-24}$ meter |
|  | $10^{-26}$ meter |
|  | $10^{-28}$ meter |
|  | $10^{-30}$ meter |
|  | $10^{-32}$ meter |
| Planck length | $10^{-34}$ meter |
|  | $10^{-36}$ meter |

I approach this overview through the above list of objects of various sizes. A person is about two meters tall, so a person is listed on the length scale of 1 meter $= 10^0$ meter. (Of course, not all people are the same size, and even if they were, two meters is not the same as one meter. But this list is just a rough guide. This table goes down to objects that are much smaller than atoms, and the basic point — that people are a whole lot bigger than atoms — is made whether people are listed as about one meter tall or about two meters tall.) The list goes to smaller and smaller lengths until it reaches microscopic objects that were not discovered until the end of the nineteenth century. There is a wide range of lengths here — a person is a million times bigger than a bacterium — but classical mechanics is able to explain phenomena at all these length scales.

But here the domain of classical mechanics ends. The structure of atoms was under intense investigation in the 1910s and 1920s, and everyone's first thought was of course to apply classical mechanics to these new length scales. Everyone did, and the results were catastrophic — classical mechanics made a number of patently incorrect predictions about atomic phenomena. Physicists first attempted to work within the framework of classical mechanics by invoking new force laws within the old framework to explain the new observations. These attempts failed. Then they tried to make the smallest possible modifications of the classical framework. Eventually these attempts failed also, and physicists were forced to develop the entire new framework of quantum mechanics to explain these facts. It took a long time growing, but once it arrived the amplitude framework, coupled with rules for assigning amplitudes, was able to explain atomic phenomena with extraordinary accuracy.

The story does not stop here, however. In the 1930s physicists probed the even smaller world of the atomic nucleus. Many strange and wonderful phenomena were uncovered. There was talk that quantum mechanics would not be able to explain these new observations, and that it would have to yield to yet another framework. But no: after sufficient thought and experimentation it was found that the amplitude framework was adequate for explaining nuclear phenomena, although new rules for assigning amplitudes had to be developed.

In the 1950s and 1960s the subnuclear world was investigated in detail. New elementary particles were discovered, new and strange interactions were found, and there was talk that a new version of mechanics would be necessary to explain all the observations. But after a while it was found that the quantal framework was perfectly adequate for the subnuclear world, once the proper rules for assigning amplitudes were uncovered. Now the nucleus is known to be made of neutrons and protons, which in turn are made up of quarks. Studies of quarks have led to measuring the shortest length ever experimentally investigated, about $10^{-19}$ meter. This length is as small, relative to an atom, as an atom is small, relative to a person. All the way down this staircase, the framework of quantum mechanics has proved to be adequate.

But while experimentalists — for now — cannot look smaller than $10^{-19}$ meter, there is nothing to stop theorists from speculating about even shorter length scales. Right now a lot of theoretical investigation centers on lengths around $10^{-35}$ meter, the so-called Planck length, where quantum effects become important for the gravitational force. The Planck length is even smaller, relative to a nucleus, than a nucleus is, relative to a person. In the 1980s theorists started to do calculations concerning phenomena at this length scale, and all sorts of impossible things started to come out. There was talk that a new framework of mechanics would be needed to replace the quantal framework, but eventually new rules for assigning amplitudes were found that enable calculations to be performed consistently. These new rules go under the name of "superstring theory", and they are very strange indeed: They predict a universe of nine spatial dimensions, six of which have curled up into little tubes so tiny that we don't notice them. (In fact, the little tubes are *so* tiny that atoms don't notice them either.) They describe a world where every particle has a complementary "sparticle", and where elementary particles themselves are more like threads or handkerchiefs than like dots. Strange as this theory is, however, its newness falls entirely within the domain of rules for assigning amplitudes — it employs exactly the same quantal framework that was uncovered in the 1920s.

In short, the framework of quantum mechanics has proven to be remarkably resilient, capable of explaining phenomena all the way from $10^{-10}$ meter to $10^{-35}$ meter. (In fact it also explains phenomena at lengths above the atomic scale, because these phenomena are governed by classical mechanics and, as we mentioned briefly in chapter 1 and will see in more detail in chapter 14, classical mechanics is nothing but an approximation to quantum mechanics that is accurate only at large length scales.) It has often happened that new amplitude rules were needed to explain the new phenomena discovered when a new length scale was investigated, but so far such new rules have always slipped seamlessly into the amplitude framework.

What of the future? We can expect that physicists will keep on investigating new phenomena. We can expect that new rules for assigning amplitudes will be uncovered. Will these new rules always fit into the by-now-familiar framework? It is of course impossible to know what will happen when these investigations are carried out, but my own guess is that the quantal framework is *not* the final word. My guess is that at some point someone will investigate a phenomenon — perhaps a newly discovered one, perhaps an old one that hadn't received the attention it deserved — and find that it cannot be fit into the quantal framework, no matter how hard scientists attempt to force it in. When that happens, a new framework will have to be developed. If you don't like quantum mechanics, this might make you happy, but watch out. It is my guess that this new framework will seem, to our classical sensibilities, even further away from common sense, even less intuitive, even stranger, than quantum mechanics.

## 10.3  References

One of the best descriptions of the amplitude framework is

> R.P. Feynman, *QED: The Strange Theory of Light and Matter* (Princeton University Press, Princeton, New Jersey, 1985) pages 36–76.

Three wonderful illustrated guides to length scales are the short film

> *Powers of Ten: A film dealing with the relative size of things in the universe and the effect of adding another zero* directed by Charles and Ray Eames (Pyramid Films, Santa Monica, California, 1978),

the associated book

> Philip and Phylis Morrison, *Powers of Ten: A book about the relative size of things in the universe and the effect of adding another zero* (Scientific American Library, San Francisco, 1982),

and the CD-ROM (which includes the film)

> *Powers of Ten Interactive* by Demetrios Eames and the Eames Office (Pyramid Media, Santa Monica, California, 1999).

The Oscar-nominated large-screen motion picture

> *Cosmic Voyage* directed by Bayley Silleck (Imax Corporation, 1996)

treats the same material in a very different way. If you ever have a chance to see this film, then do so, you won't regret it. A scholarly and historical (but still very readable) survey of length scales is

> Abraham Pais, *Inward Bound: Of Matter and Forces in the Physical World* (Clarendon Press, Oxford, UK, 1986).

The deepest experimental plunge into smallness — the investigation at the scale of $10^{-19}$ meter mentioned in this chapter — is described in

> F. Abe *et al.*, "Measurement of dijet angular distributions by the collider detector at Fermilab", *Physical Review Letters*, **77** (1996) 5336–5341.

Superstring theory is treated at the popular level by

> Michio Kaku, *Hyperspace* (Oxford University Press, New York, 1994),
> Timothy Ferris, *The Whole Shebang: A State-of-the-Universe(s) Report* (Simon and Schuster, New York, 1997),
> Michael J. Duff, "The theory formerly known as strings", *Scientific American*, **278** (February 1998) 64–69.

## 10.4 Problems

10.1 *Barriers to understanding.* (Compare problem 4.11.) Distinguish between "a description of quantum mechanics", "an understanding of quantum mechanics", and "an explanation for quantum mechanics".

10.2 *Logical contradiction vs. unfamiliar visualization.* For the magnetic needle of a silver atom, we found that

> If *the atom's magnetic needle were just like a classical arrow*, then *the conundrum of projections* would be much worse than a puzzle, it would be a logical contradiction. We are able to regain logical consistency only by abandoning the mental picture of *a magnetic needle as a pointy stick*.

Change the three phrases in italics to produce a parallel statement concerning the position of an atom.

10.3 *States of observed atoms.* Demonstrate that in figure 10.1 we cannot give an amplitude for the atom to move from one place to another, but we must instead give an amplitude for the atom and the photon to move from their two initial positions to their two final positions.

# 11

## Working with Amplitudes

The first section of this chapter shows that the mathematical representation of amplitude cannot be as simple as a real number, but must be at least as complicated as a two-dimensional arrow. If you're willing to accept this as fact, then you may skip that rather technical and involved section. But in no case should you skip over the second section of this chapter, which makes a simple but subtle and important general point.

### 11.1   Amplitude is represented by an arrow

I'm going to introduce one more type of analyzer: the "front–back analyzer" (also called the "$y$ analyzer"). This will be the last new analyzer, I promise.  The left half of this analyzer is just like the left half of a traditional Stern–Gerlach analyzer, with its traditional non-uniform magnetic field. But while the right half of the traditional Stern–Gerlach analyzer contains only plumbing to make sure the atoms come out parallel to the sides of the box, the right half of the front–back analyzer contains also a magnetic field that changes direction slowly from place to place. Along the path towards the upper exit, the magnetic field starts by pointing straight up. A little farther on it tilts a bit to the right. The tilt angle of the field increases gradually until, just before the exit, the field points directly to the right. The path towards the lower exit is similar, except that in this case the field starts out pointing down and gradually tilts until it points directly to the left.

How does this tilting field affect a passing atom? Only experiment can tell for sure, but the following arguments are suggestive and turn out to give the correct answer. An atom that leaves the left half of the front–back analyzer through the upper branch has $m_z = +m_B$, that is, its magnetic arrow is "more-or-less pointing up". (I use the qualifier "more-or-less" just to remind you that atomic magnetic arrows don't point in the same

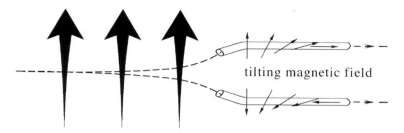

non-uniform magnetic field

definite manner that sticks do.) So when it encounters the tilting magnetic field, the field is pointing in the same direction as the magnetic arrow. It seems reasonable that, as the atom gradually makes its way through the corridor of tilting field, the atom's magnetic arrow will be dragged right along with the field. Thus when the atom leaves the upper exit its arrow points directly to the right. In other words, an atom leaving the upper exit leaves with a definite value for the projection of its magnetic arrow on the $y$ axis, namely $m_y = +m_B$. (Note that this atom has a definite value of $m_y$, so it no longer has a definite value of $m_z$ or $m_x$.) Similarly, an atom leaving the lower exits leaves with $m_y = -m_B$. As before, we package this apparatus up into a box inscribed with a distinctive symbol.

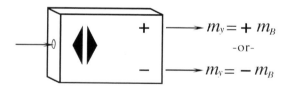

*Repeated measurement experiments with the front–back analyzer*

*Experiment 11.A.1.* Measurement of $m_y$, then $m_y$ again.

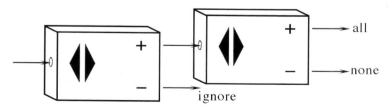

This experiment behaves exactly like the repeated measurement experiment 4.1 on page 23. An atom that leaves the + exit of the first analyzer (i.e. one with $m_y = +m_B$ leaving the first analyzer) will always leave the

+ exit of the second analyzer (i.e. it still has $m_y = +m_B$ when entering the second analyzer). This experiment just confirms a very reasonable expectation.

*Experiment 11.A.2.* Measurement of $m_y$, then $m_y$ with a tilted front–back analyzer.

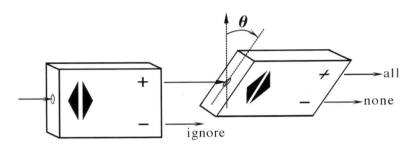

An atom found to have $m_y = +m_B$ at the first analyzer is found to have $m_y = +m_B$ at the second analyzer, regardless of the orientation angle $\theta$. This is reasonable because tilting the front–back analyzer doesn't change the character of the output atoms: their magnetic arrows are "more-or-less" pointing front or back, not up or down, so when the analyzer is tilted they're still pointing front or back.

*Experiment 11.A.3.* Measurement of $m_z$, then $m_y$ with a tilted front–back analyzer.

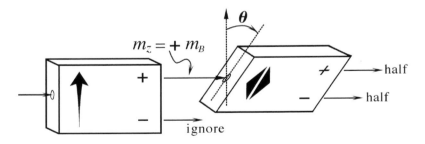

We still expect that tilting the front–back analyzer will have no effect. In other words, we still expect that the statistics of exit from the second analyzer will be independent of the orientation angle $\theta$. Furthermore, because the direction "straight up" bears the same relation to the direction "directly right" as it does to the direction "directly left" you might expect that an atom with $m_z = +m_B$ will have the same relation to an atom with $m_y = +m_B$ as it does to an atom with $m_y = -m_B$. Experiments show both of these expectations to be correct: The statistics of exit from the second analyzer are that half leave the + exit and half leave the − exit, regardless of the angle $\theta$.

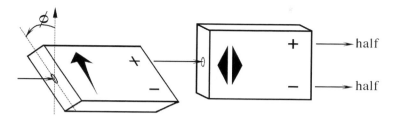

Of course, tilting the second analyzer to the right by 17° is equivalent to tilting the first analyzer to the left by 17°. We conclude that if an atom has a definite value for the projection of its magnetic arrow on any axis in the $(x, z)$ plane (that is, an atom in any of the states discussed before this chapter began: states like $m_z = +m_B$, $m_z = -m_B$, $m_{(-x)} = +m_B$, or $m_{39°} = -m_B$) and if the value of $m_y$ is measured, then the chances are half-and-half that the atom will be found to have $m_y = +m_B$ or to have $m_y = -m_B$.

### Interference experiments with the front–back analyzer

We can make an interferometer from a front–back analyzer just as we did from a Stern–Gerlach analyzer.

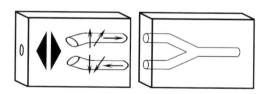

I will describe several experiments using the apparatus sketched below. In all cases the input atom has $m_z = +m_B$. The atom passes through a vertical front–back interferometer, and then passes into a regular Stern–Gerlach analyzer (not a front–back analyzer) tilted at an angle $\theta$ relative to the vertical. An atom leaving the + exit of this analyzer (in which case it has $m_\theta = +m_B$) is considered output; an atom leaving the − exit is ignored. The atom is not watched at either of the branches.

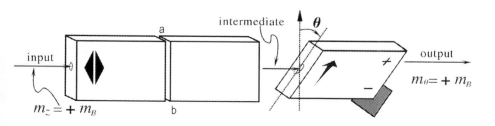

*Experiment 11.B.1.* Branch a is blocked.

The probability of passing from input to intermediate is $\frac{1}{2}$.
The intermediate atom has $m_y = -m_B$.
The probability of passing from intermediate to output is $\frac{1}{2}$.
The overall probability of passing from input to output is $\frac{1}{2} \times \frac{1}{2} = \frac{1}{4}$.

*Experiment 11.B.2.* Branch b is blocked.

This is the same as experiment 11.B.1 except that the intermediate atom has $m_y = +m_B$.

*Experiment 11.B.3.* Neither branch is blocked.

The probability of passing from input to intermediate is 1.
The intermediate atom has $m_z = +m_B$.
The probability of passing from intermediate to output is $\cos^2(\theta/2)$.
    (See figure 4.1 on page 27.)
The overall probability of passing from input to output is $\cos^2(\theta/2)$.

Given the results of these three experiments, we attempt to assign amplitude arrows to the two paths "input to output through branch a" and "input to output through branch b". The amplitude arrow assigned to "input to output through either branch" will be the sum of these two arrows. We don't know the orientations of the arrows, but we do know that the magnitudes of the three arrows must be $\frac{1}{2}$, $\frac{1}{2}$, and $\cos(\theta/2)$ respectively.

Now, it is entirely possible (as demonstrated in the figure below) to find two arrows of magnitude $\frac{1}{2}$ and $\frac{1}{2}$ that add up to produce a sum arrow of magnitude $\cos(\theta/2)$ for any angle $\theta$.

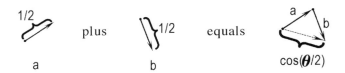

But it is quite impossible to find two real numbers of magnitude $\frac{1}{2}$ (that is, either $+\frac{1}{2}$ or $-\frac{1}{2}$) that add up to produce a number continuously varying with angle $\theta$: these numbers must add up to either 0 or 1.

We conclude that whatever mathematical entity is used to represent an amplitude, it must be at least as complicated as a two-dimensional arrow. Of course, it might be even more complicated: for example an arrow in three dimensions. But as far as anyone knows, two-dimensional arrows are sufficient.

## 11.2    Amplitudes for the Einstein–Podolsky–Rosen experiment

This section is much shorter and much less technical than the previous section, but the result is more important. Whenever I have discussed amplitudes, I have been careful to associate an amplitude with an action (also called "a process") rather than with a particle. For example, I would talk about "the amplitude to go from input to output through branch a" and never "the amplitude the particle has if it went through branch a". The latter phrase, I am sure you realize, contains a misimpression about the nature of quantal interference (see page 78). However, every example I have given so far involves a single particle, so despite my care it is easy to get the mistaken impression that an amplitude arrow must be associated with a specific particle, and that it acts somehow like an arrow hanging off of that particle. This section gives an example in which the action involves a *pair* of particles, showing concretely that amplitudes are not associated with individual particles.[*]

Recall the first Einstein–Podolsky–Rosen experiment, described in section 6.1 (page 40) and represented on the next page by figure 11.1. The initial condition is given by state A in the figure. Possible final states are given by states B, C, and D. Remember from section 6.1 that the two atoms always leave their respective analyzers from opposite exits. In terms of the figure, this means that there is some amplitude for going from state A to state B, and some amplitude for going from state A to state C, but there is no amplitude for going from state A to state D.

Now, look at this from the perspective of the atom released from the source and flying to the right towards its detector. If it were in a state like $m_x = +m_B$, then it would have some amplitude to leave its detector through the $+$ exit and some amplitude to leave its detector through the $-$ exit. Similarly for the atom flying to the left. If we assigned an amplitude to each of the individual particles in the manner suggested, then it would be impossible to prevent the system from ending up in state D of the figure. But in fact the system never does end up in state D. We conclude that one cannot assign one amplitude to an act performed by the atom on the right and a second amplitude to an act performed by the atom on the left. Instead, we must assign a single amplitude to an action by the pair of atoms.

When the two atoms are flying from the source to their analyzers, it is not possible to assign each one to a state like $m_x = +m_B$ or $m_x = -m_B$. Instead the two particles together must be assigned to a single state. Such states are called entangled states. This is an excellent name,[†] because

---

[*] In technical terms, this example shows that a wavefunction is a function in configuration space, not position space.

[†] It was coined by Schrödinger in 1935.

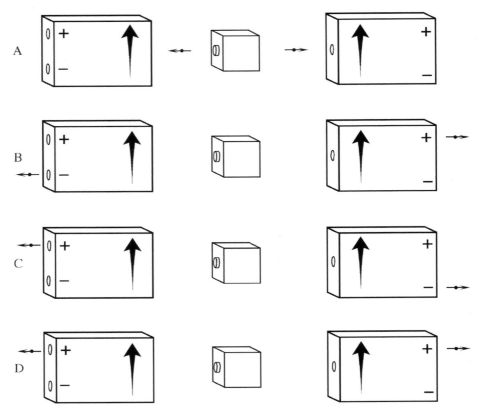

Fig. 11.1.  Various states for two atoms in the first Einstein–Podolsky–Rosen experiment.

it suggests quite graphically (and quite correctly) that what happens to one particle is mixed up with what happens to the other. Entangled states come up not only in abstruse discussions on the foundations of quantum mechanics, but also in the practical day-to-day work of atomic and molecular physics. If entangled states were to go away, so would most of chemistry.

## 11.3    Problems

11.1  *Other schemes for amplitudes.* Mr. Parker is uncomfortable with the idea that amplitudes must be represented by two-dimensional arrows. He uses the symbol $A_a$ to represent "the amplitude to pass through branch a", the symbol $A_b$ to represent "the amplitude to pass through branch b", and the symbol $A_{a,b}$ to represent "the amplitude to pass through both branches".

(a) "I know that we want to have a mathematical representation for amplitude in which

$$A_{a,b} = A_a + A_b,$$

and I know that we must sometimes have two non-zero amplitudes summing to zero. But why can't we represent amplitudes by real numbers and assume that the probability is the absolute value of the amplitude rather than the square of the amplitude?" Convince Mr. Parker that no such scheme is consistent with the facts outlined in section 11.1.

(b) "All right, you've convinced me," says Mr. Parker. "But what about a scheme in which

$$A_{a,b} = \sqrt{(A_a)^2 + (A_b)^2}$$

which also ensures that probabilities are always positive?"

11.2 *Magnitudes of amplitude arrows.* Find the magnitude of the amplitude arrow associated with going from state A to state B in figure 11.1. Similarly for going from state A to state C and from state A to state D. Do not attempt to find the directions of these arrows.

11.3 *Distant measurements.* "I've got it now!" says Mr. Parker. "I was wrong back in problem 6.2 when I suggested that the two atoms in experiment 6.1 were produced in the states $m_x = +m_B$ and $m_x = -m_B$. But now I see that they were produced in the states $m_y = +m_B$ and $m_y = -m_B$. That explains all the observations!" Show that Mr. Parker's new suggestion is still not consistent with the observation in experiment 6.1 that the two atoms always leave through exits of the opposite sign.

11.4 *What if they weren't entangled?* Suppose that, in figure 11.1, the atom on the right had probability $\frac{1}{2}$ of leaving either the + or the − exit of its analyzer, and similarly for the atom on the left. (This supposition is correct). Suppose also that the actions of the two atoms were not entangled. (That is, the actions were uncorrelated — this supposition is not correct.) Under these assumptions, what would be the probability of beginning in state A and ending in state D?

11.5 *Measurement and entangled states.* Interpret the measurement experiments of figure 10.1 (page 80) in terms of entangled states. In particular, show that it is not possible to assign one amplitude for the exit taken by the atom and a second amplitude for the final position of the photon. Instead, one must use a single amplitude to describe both the atom and the photon.

# 12

# Two-Slit Inventions

In chapter 9 we concluded that in quantal interference experiments a single atom passes through both branches of an interferometer. In chapter 10.1 we firmed up that everyday-language expression to the technical phrase "there is an amplitude for the atom to go through either branch". Exactly what do these strange statements mean? How can our minds grow familiar with a real quantal atom, which behaves so unlike a small, hard marble? To prepare for these questions, this chapter examines two variations of the quantal interference experiment. This chapter is not absolutely essential for the logical development of the book, but it dramatically underscores that quantal interference demands a total rethinking of our picture of the atom — no simple trick will suffice.

## 12.1   The Aharonov–Bohm effect

It's possible to build a box called a "corkscrew" from a uniform magnetic field twisted into one turn of a spiral (see figure below). At the left edge of the box the magnetic field points straight out of the page (that is, in the $+x$ direction). Moving towards the right the magnetic field slowly dips until it points straight down, then continues to twist until it points into the page, then straight up, until finally — at the right edge of the box — the magnetic field again points straight out of the page. (The magnetic field always points perpendicular to the $y$ direction.)

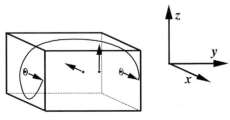

How do atoms behave when they pass through a corkscrew? Only experiment can tell for sure, but the following argument is suggestive and turns out to be correct. If an atom with $m_x = +m_B$ enters a corkscrew, it enters with its magnetic arrow pointing "more-or-less" in the same direction as the magnetic field. (The qualifier "more-or-less" is there just to remind you that atomic magnetic arrows don't point in the same definite manner that sticks do.) It seems reasonable that such an atom's arrow would be dragged around by the field as the atom passes down the center of the corkscrew, and thus that it will emerge with its magnetic arrow still pointing in the $+x$ direction, after having executed a complete flip. This expectation is confirmed by experiment: If an atom with $m_x = +m_B$ enters a corkscrew, it emerges with $m_x = +m_B$ and no experiment performed on that single atom can tell whether it passed through a corkscrew or through an empty box. As far as an atom with $m_x = +m_B$ is concerned, passing through a corkscrew is equivalent to doing nothing.

Using a corkscrew we can turn the interference experiment sketched on page 65 into something surprisingly different (see figure below). Recall from experiment 9.3 on page 66 that if an atom with $m_z = +m_B$ enters an unblocked interferometer, it leaves in the same state, namely with $m_z = +m_B$. The two halves of the interferometer can be drawn apart and experiment repeated: the results are exactly the same. Now, insert a corkscrew into branch a so that any atom passing through branch a also passes through the corkscrew. Remember that an atom passing through branch a has $m_x = +m_B$ and that for such atoms a corkscrew does nothing. Experiment reveals that after the corkscrew is added any atom entering with $m_z = +m_B$ emerges in a *different* state, namely with $m_z = -m_B$.

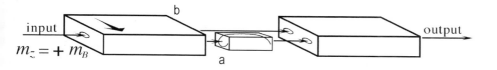

How can this be? Didn't we say, just two paragraphs ago, that "as far as an atom with $m_x = +m_B$ is concerned, passing through a corkscrew is equivalent to doing nothing"? Indeed we did. But the atom *doesn't* simply pass through a corkscrew — it passes through *both* branches.

If an atom went through branch a, or through branch b, then inserting a corkscrew in branch a would have no effect on the interference experiment. The fact that the corkscrew *does* have an effect proves that a single atom goes through both branches.

## 12.2   Delayed choice experiments

In our primary quantal interference experiment (experiment 9.3 on
page 66), the interferometer was horizontal and the trailing analyzer
was vertical. In this situation each unwatched atom "goes through both
branches" of the horizontal interferometer and emerges from the + exit
of the trailing analyzer. In contrast, each watched atom goes through only
one branch (whichever branch it is observed to take), and has probability
$\frac{1}{2}$ of emerging from the + exit.

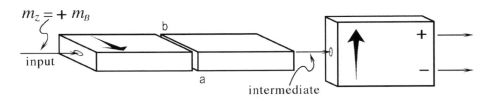

A variation on this experiment is to orient the trailing analyzer horizontally. In this situation both watched *and* unwatched atoms have
probability $\frac{1}{2}$ of emerging from the + exit of the trailing analyzer. An
atom observed to pass through branch a always emerges from the +
exit, so it is tempting (although wrong) to believe that an unobserved
atom found emerging from the + exit had also passed through the single
branch a. Indeed, if this were the only experiment we ever performed, we
would never have to deal with ideas like "a single atom goes through both
branches" or "a watched atom behaves differently from an unwatched
atom" — we could always be content with each atom taking a single
definite path through the apparatus.

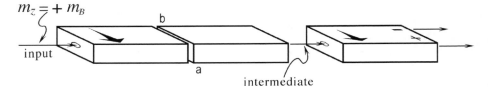

Is it possible, then, that the atom goes through both branches when
the trailing analyzer is vertical but goes through a single branch when
the trailing analyzer is horizontal? This possibility is called a "conspiracy
theory" because the atom somehow senses the arrangement of the apparatus and behaves accordingly. (In fact, it senses the arrangement of the
trailing analyzer even as it passes through the interferometer, having not
yet encountered the trailing analyzer!) Quantum mechanics, in contrast,
predicts that an unwatched atom goes through both branches in either

case. One way to test the two alternatives is through a "delayed choice experiment". In this experiment the trailing analyzer is mounted on a pivot and can be swung from vertical to horizontal at a moment's notice.

Suppose an atom enters the interferometer while the trailing analyzer is horizontal. Then, according to the conspiracy theory, it goes through one branch or the other and emerges with $m_x = +m_B$ or with $m_x = -m_B$. Now, as the atom flies from the interferometer to the trailing analyzer, the trailing analyzer is quickly swung to the vertical position. When an atom with either $m_x = +m_B$ or $m_x = -m_B$ enters a vertical analyzer, it has probability $\frac{1}{2}$ of emerging from the $-$ exit. Thus the conspiracy theory predicts that half of such atoms will leave through the $-$ exit of the now-vertical analyzer. Quantum mechanics predicts that, regardless of the orientation of the trailing analyzer, each atom goes through both branches, each atom flying from the interferometer to the analyzer has $m_z = +m_B$, and thus all such atoms will leave through the $+$ exit of the vertical analyzer.

This conceptual experiment has been realized in several different ways — each time with somewhat different details — and quantum mechanics has been confirmed every time.

## 12.3   References

The Aharonov–Bohm effect was predicted from quantum theory in 1959 in a form and context very different from the one described here, and this prediction gave birth to a whole series of experiments and arguments. This story is told in the highly technical yet very beautiful little book

M. Peshkin and A. Tonomura, *The Aharonov–Bohm Effect* (Springer–Verlag, Berlin, 1989).

No such overview has been written for delayed choice experiments. The closest approach is

George Greenstein and A.G. Zajonc, *The Quantum Challenge* (Jones and Bartlett, Sudbury, Massachusetts, 1997), pages 37–42.

Recent developments are reported in

T. Kawai *et al.*, "Development of cold neutron pulser for delayed choice experiment", *Physica* B, **241** (1998) 133–135.

# 13

# Quantum Cryptography

Quantum mechanics is valuable because it opens a discussion about the nature of reality, because it demonstrates the power of reason in revealing the truth even when common sense is an obstacle, and simply because it is good to know how our universe ticks ("knowledge is better than ignorance"). But it is also valuable because a host of practical devices, from lasers to transistors to superconductors, all work because of quantum mechanics.

Most of these applications are beyond the scope of this book. I could tell you in vague terms how a laser works, but I could never convince you that my description was correct — you would have to accept it on my authority, and acceptance on the basis of authority is the very antithesis of scientific thought. However, there is one very recent, very exciting application of quantum mechanics that can be treated in full within the "rigorous but not technical" style of this book, namely the use of quantum mechanics to send coded messages. (You may skip this chapter without interrupting the flow of the book's argument.)

## 13.1  Can you keep a secret?

Sending coded messages is a part of life. Governments and businesses need to transmit secrets that would be deadly in the wrong hands (military plans, formulas for explosives, etc.). But even you have information that you don't want everyone in the world to know: your bank balance, your voting record, your vacation plans. I'm not suggesting that you should be embarrassed about your bank balance, but it's your private information and no eavesdropper has any right to it. For this reason, when credit card and automatic teller machine transactions are sent over public telephone lines, the messages are sent in code. Such codes are enormously valuable, and there is an ongoing policy debate about who can use the best codes. The United States Commerce Department classifies difficult-to-break codes as munitions (along with guns and bombs and fighter jets) and prohibits their export from the country.

Cryptography is the art of sending information from place to place in coded form so that it will be meaningless to any eavesdropper who might intercept it. The problem of cryptography is to find a mechanism for one person — conventionally named "Alice" — to send secret information to another person — "Bob" — while a third person — "Eve" — might or might not be eavesdropping. A number of coding schemes are in use, but I will describe only one, the "Vernam cipher" or "one-time pad scheme", because it is the only coding scheme that has been proven to provide perfect, unbreakable security.

Suppose Alice wants to send computer mail to Bob. Computers store information internally as clusters of ones and zeros, each digit called a "bit". In the standard representation of characters by bit clusters — used worldwide by nearly all computers — each character of text is represented by a cluster of seven bits. For example, the letter "a" is represented by the cluster "1100001", the numeral "4" by the cluster "0110100", the comma by the cluster "0101100", and a blank space by the cluster "0100000". Thus whenever Alice sends computer mail to Bob, she is sending him a long string of bits — ones and zeros — which his computer can easily interpret as letters and numbers. Unfortunately, Eve's computer can do so just as easily.

To maintain secrecy, before sending her message Alice produces a random string of bits — called the "key" — exactly as long as her message. She then encodes each bit of her message according to the key: If the fifth bit of the key is a zero, then she sends the fifth bit of her message unaltered, but if the fifth bit of the key is a one, then she reverses the fifth bit of her message (if it is a 1, she sends a 0; if it is a 0, she sends a 1). For example, if Alice encoded the character "a" using the key 0101101, the resulting coded message would be 1001100 as shown here:

> 1100001   (standard representation for "a")
> 0101101   (key)
> 1001100   (code for "a")

After encoding her message, Alice sends Bob not only her coded message, but also her key. Bob decodes Alice's message in the same way that she encoded it: He preserves the message bits corresponding to zeros in the key, and alters the message bits corresponding to ones in the key. Bob's decoded message is then exactly the same as the one that Alice started with.

If Eve intercepts only the coded message, it won't do her any good. Of course, the coded string of bits will translate to *some* message in the standard representation, but that message will be gibberish. Eve might try to decode a 91-bit message with every possible 91-bit key, but that won't help her because she would then produce every possible 91-bit statement, including

```
"Withdraw $100",
"Buy stock now",
"I love Bob!!!",
"I despise Bob",
"Bomb Baghdad.",
```

and a great many statements like

```
"U&87{{ ^(aqNq".
```

However, if Eve intercepts both the coded message *and* the key, then Eve can decode the message just as easily as Bob can. The key must instead be transmitted through some separate secure channel that Eve cannot intercept. But if Alice and Bob have a secure channel, they don't need to bother with codes at all! Alice and Bob might hire a courier (who is a secure channel) to deliver several identical keys to both Alice and Bob at the beginning of each week, and they can use those keys throughout the week. But then Eve might bribe the courier. (It doesn't work to use one key over and over — there are easy ways to break the code if the same key is used even twice. This is the origin of the name "one-time pad".)

In short, the problem with the Vernam cipher is not the distribution of *messages* but the distribution of *keys*. It is ironic but nevertheless true that an important problem for contemporary business and government is the generation and distribution of random numbers.

### 13.2    Distributing random keys

Since probability and randomness are intrinsic to quantum mechanics, you might guess that quantum mechanics could provide some help with the problem, and indeed it does. Suppose Alice and Bob set up experiment 6.1, "EPR distant measurements" (page 40) with one vertical analyzer next to Alice, the other next to Bob, and the source of atoms between them. They set the source to automatically generate pair after pair of atoms, and when those atoms reach their analyzers Alice and Bob both record the exits taken. If Alice records "+ + − + −", then Bob records exactly the opposite pattern, namely "− − + − +". Alice turns her readings into a cryptographic key by converting each + to a 1 and each − to a 0. Bob does the same with the opposite convention, namely + goes to 0 and −

goes to 1. Now both Alice and Bob have the same random key and can send a coded message using the Vernam cipher.

Unfortunately, Eve can easily break into this system by inserting a vertical interferometer between the source and Bob. Eve watches each atom pass through her interferometer. When one goes through her top branch, she knows that Bob will get a + and Alice will get a −. Similarly for her bottom branch. Eve gets the key, Eve breaks the code.

## 13.3  Distributing random keys securely

To prevent eavesdropping, Alice and Bob instead set up experiment 6.2, "EPR random distant measurements" (page 42) with one randomly tilting analyzer next to Alice, the other next to Bob, and the source set to "automatic" as before. When an atom reaches an analyzer, Alice (or Bob) records both the analyzer orientation (**A**, **B**, or **C**) and the exit taken (+ or −). Recall that if the two analyzers are set to the same orientation, then the two atoms emerge from opposite exits, but if they are set to different orientations, then the two atoms might emerge from either similar or opposite exits. (They emerge from the similar exits with probability $\frac{3}{4}$ and from opposite exits with probability $\frac{1}{4}$, but this fact is not needed in what follows.)

Alice and Bob run this experiment for a long time, and then send to each other the list of their analyzer orientations. (Each list looks something like **BBACABBC**.... There's no need to encode these messages: if Eve intercepts them, the lists won't help her.) When they compare lists Alice and Bob find that in most cases their two detectors were set to different orientations, but in about one-third of the cases the detectors happened to have the same orientation. They discard the exit information (the +s and −s) for those cases with different orientations, and use the cases with the same orientation to construct a key just as they did previously. Now that they have identical keys, Alice and Bob can send coded messages using the Vernam cipher.

What if the nefarious Eve tries to intercept the key in this distribution scheme, as she did previously? Suppose Eve places a vertical interferometer between the source and Bob, and watches each passing atom to see which branch it takes. (For definiteness, assume that Bob and Alice are equally distant from the source.) Now when atoms arrive at the tilting analyzers used by Alice and Bob, they are no longer in an entangled state: instead, one atom has $m_z = +m_B$ and the other has $m_z = -m_B$. If the two detectors are vertical (orientation **A**) this makes no difference: the two atoms still emerge from opposite exits. But if both detectors are in orientation **B**, then there is some probability (it turns out to be $\frac{3}{8}$) that the two atoms will emerge from the same exit. Alice and Bob therefore agree beforehand that they will not use the entire key as generated above. Instead Bob will mail, say, the first half of his key back to Alice. If Bob's first half matches Alice's first half, then Alice knows that no one was eavesdropping on the key distribution and that it is safe to send her message coded using the second half of the key. If the two half-keys don't match, then Alice doesn't send her message on to Bob but instead calls the police and tells them to search for Eve.

This precise method of key distribution is not practical: it relies on a source of atoms that just happens to be conveniently placed between Alice and Bob, it involves sending a lot of information back and forth that is ultimately ignored, and in the end it doesn't actually ensure privacy, it merely lets the legitimate users know whether or not someone is listening in. There are other quantum cryptography schemes that lack many of these drawbacks, and these are so promising that they have raised the interest even of commercial communication companies. (The experiment of Nicolas Gisin mentioned on page 42 was supported in part by Swiss Telecom.) Quantum cryptography is a new field (the first experiment was performed in 1989) but both theory and practice are growing rapidly and hold the promise of practical applications from the most esoteric fundamentals of quantum mechanics.

## 13.4    References

Charles H. Bennett, Gilles Brassard, and Artur K. Ekert, "Quantum cryptography", *Scientific American*, **267** (4) (October 1992) 50–57.

Wolfgang Tittel, Grégoire Ribordy, and Nicolas Gisin, "Quantum cryptography", *Physics World*, **11** (3) (March 1998) 41–45.

# 14

# Quantum Mechanics of a Bouncing Ball

We started to investigate quantum mechanics by considering only the quantization of magnetic arrows. In our explorations we found out that the magnetic arrow had some funny properties (for example, it was possible that $m_x$ did not have a definite value), but at first it seemed that other properties, such as the position of an atom, behaved in the familiar classical way. Eventually (section 9.3) we found that it was also possible to have an atom without a definite value for its position. In this chapter, we investigate what happens when we apply quantum mechanics to a particle's position.

## 14.1   Ball bouncing from a floor

This chapter will show our framework for quantum mechanics in action, by applying it to the problem of a ball bouncing from a floor. Let us use a very fast ball, such as an electron, so that we can ignore the force of gravity. (We restrict ourselves to an electron that is moving fast on a human scale but slow compared to the speed of light, so that relativistic considerations don't come into play. Also, the magnetic arrow associated with the electron has no effect on the phenomena described in this chapter, so I won't mention it again.)

   Imagine a source of balls that could send a ball flying in any direction, for example a hot tungsten filament that boils off electrons. Suppose a ball begins at point P, bounces off the floor, and ends up at point Q. (Points P and Q are equally distant from the floor. To make sure that the ball bounces off the floor rather than goes directly from P to Q, we put a barrier half-way between the two points.) In classical mechanics this sequence can happen in only one way, namely by the ball hitting the floor midway between points P and Q.

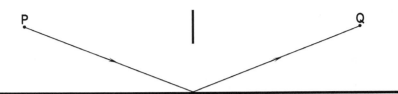

But in quantum mechanics there are many ways of going from P to Q, and each way will contribute some amplitude to the process.* In this discussion we will ignore curlicue paths from P to Q and consider only those paths that consist of a straight line from P to some point on the floor, and then a straight line from that point to Q. We will also ignore paths that go out of the plane of the page. Although it's not yet obvious (see problem 14.2), these simplifications do in fact give the correct answer, and they also give a correct feel for how to do the full problem! Here are the steps (from page 78) in finding the probability to pass from P to Q.

*Step 1: Enumerate all the paths from* P *to* Q. The complete enumeration is difficult, because there are an infinite number of paths, so I'll just draw some representative paths and label them according to the point where they hit the floor.

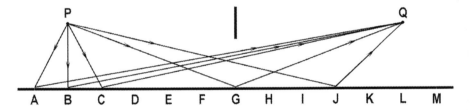

The classical path is the one that hits the floor at point G, but in quantum mechanics we must consider *all* possible paths.

*Step 2: Assign an amplitude arrow to each path.* Here we need to be creative. You might think that the classical path is the "most important", and thus should be assigned the longest arrow. But this is not the case at all. The correct rule assigns to every path an arrow of the same magnitude, but with the different arrows pointing in different directions. The arrow assigned to a path is found by starting with an arrow pointing directly to the right and then rotating it counterclockwise according to the length of the path from P to Q:

$$\text{number of rotations} = 1.51 \times 10^{26} \times \text{(length of path in cm)}$$
$$\times \text{(mass of ball in grams)} \times \text{(speed of ball in cm/s)}.$$

---

* In both classical and quantum mechanics, not all the balls starting out at point P go to point Q. None of the balls that start out simply vanish, but many of them do not go to point Q.

There is no way that you — or anyone else for that matter — can derive this rule. It is one of the fundamental laws of nature and cannot be derived from anything simpler.

> *Technical aside:* Throughout this book I have tried to be non-technical yet completely honest and truthful. In the above formula I have had to retreat somewhat from my principled stance. It is true only for particular circumstances, and I don't know how to describe these circumstances precisely without invoking technical terms. The formula's limitation involves the fact that it purports to give the amplitude for moving from point P to point Q, whereas what's really needed is the amplitude for moving from point P at time $t_p$ to point Q at time $t_q$. A symptom that the formula suffers from illness is that it invokes a speed for moving between two positions and, as we will see in section 14.3, a ball cannot have a definite position and a definite speed at the same time. While I'm off on a technical aside, let me point out that the formula above is called the "de Broglie relation", and the number $1.51 \times 10^{26}$ which appears in the formula is called the inverse of Planck's constant $h$.

The number of rotations may, of course, be a fraction. For example, 13.5 rotations would result in an arrow pointing to the left, while 182.75 rotations would result in an arrow pointing downward.

Since the paths have a variety of different lengths their associated arrows point in a variety of different directions. Figure 14.1 shows how the length varies for different paths, and the arrow below each representative path shows the amplitude arrow assigned to that path.

Notice that path A is considerably longer than path B so the arrow associated with path A has rotated much more than has the arrow associated with path B. However, path F is only a bit longer than path G, so the associated arrows are nearly parallel.

*Step 3: Add up all the arrows.* This seems like a formidable task, because we have to add up an infinite number of arrows! We set about doing it using the tried-and-true scheme of "divide and conquer". We will first add up the arrows over bundles of nearby paths, and only then will we find the grand total by adding up the sums associated with each bundle.

Consider a bundle of paths like A and B and C, where the arrow changes direction dramatically from one path to another. The arrows assigned to individual paths point first up, then down, now right, now left, so that when they are added together their sum hardly amounts to anything. But now consider the bundle of paths F and G and H. Here the arrows are nearly parallel, so they add together rather than cancel out. You can see

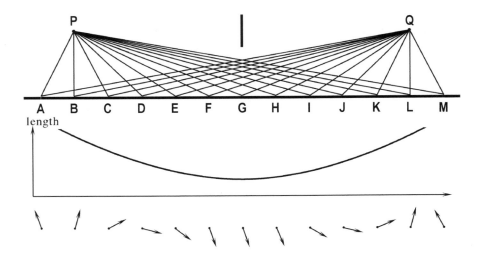

Fig. 14.1.  Representative paths from P to Q, their lengths, and the amplitude arrow associated with each.

that the grand total amplitude comes almost entirely from the bundle of paths near the midpoint G, and that all the rest of the bundles contribute very little: the corresponding pieces of floor might as well be chopped up and tossed out the door. This is precisely in accord with everyday experience, from which we know that only the midpoint of the floor is needed to bounce a ball from P to Q.

## 14.2   Ball bouncing from a floor with holes

So you see that quantum mechanics confirms your classical expectation that the ball hits the floor only at the midpoint. It is, however, a hollow victory to work so hard to obtain a result known to every child. Can we salvage anything new or surprising from this discussion? Indeed we can.

Let us chop up and toss out the right hand three-quarters of the floor, leaving only the part near points A, B, and C. In classical mechanics it is impossible to bounce a ball from point P to point Q using this remaining piece of lumber, but in quantum mechanics we can trick the ball into bouncing this way! Remember that the total arrow associated with the bundle of paths encompassing A and B and C is nearly zero because the individual arrows are pointing every-which-way: many tilt towards the right, but just as many tilt towards the left. The trick is to remove those parts of the floor responsible for arrows that tilt towards the left. What remains will be a series of slats rather than a solid floor, but the paths bouncing off the slats all have rightward tending arrows. There will be fewer paths from P to Q, but the arrows associated with those paths, instead of cancelling out, will instead add together cooperatively to produce a substantial total amplitude arrow, and hence a substantial probability of bouncing from P to Q. (It may seem strange to get more bouncing from less floor, but I suspect that by now nothing can shock you.)

If you examine this scheme quantitatively you will find that the slats must be separated by distances of about $10^{-8}$ cm. It is quite difficult to mechanically produce such closely spaced slats, but fortunately nature has provided exactly the desired bouncer: it is a crystal. The rows of atoms in a crystal act as bouncing slats, while the gaps between them act as the spaces. The bouncing of electrons off a crystal (technically called "electron diffraction") was first observed by Clinton Davisson and Lester Germer in 1927.

## 14.3  Wave-particle duality

We have seen, in some detail now, how balls behave in quantum mechanics, and you know that this behavior is utterly unlike the behavior of classical baseballs and marbles. Just as a magnetic arrow with a definite value of $m_z$ does not have a definite value of $m_x$, so an electron between release and detection does not have a definite value for its position. This means exactly what it says: it does not mean that the electron has a definite position which is changing rapidly and unpredictably, nor does it mean that the electron has a definite position but that we don't know what it is. It means that the electron just doesn't have a position, in exactly the same way that love doesn't have a color.

We found in chapter 4, "The conundrum of projections", that an atom's magnetic arrow could have a definite value for the projection $m_z$ or a definite value for the projection $m_x$ but not definite values for both at the same time. A similar statement turns out to be true for an electron: it can have a definite position or it can have a definite speed, but it cannot have both a definite position and a definite speed. There is no way for you to derive this — I'm just telling you. In fact, it is a technical detail that I ordinarily wouldn't mention to a general audience at all, but this detail has taken hold of the public imagination so effectively that many believe it to be the central, or perhaps even the only, principle of quantum mechanics. This detail is called the "Heisenberg uncertainty principle". (The term "uncertainty" actually reinforces the misconception that an electron has a definite position and a definite speed, but that we are not sure what they are. For this reason, the principle is more accurately called the "Heisenberg indeterminacy principle".)

The amplitude arrow picture first came up in association with waves, yet in quantum mechanics it describes the motion of a particle. This combination is sometimes called "wave-particle duality" or by saying "in quantum mechanics, an electron behaves sometimes like a wave and sometimes like a particle". I find such phrases unhelpful and extremely distasteful. From the world of everyday observation, we know about several classes of entities: marbles, putty balls, pond ripples, ocean breakers, clouds, sticks, balloons, etc. To insist that quantal entities must fall into one or another of these categories is utterly parochial. It is like a man born and raised in England who knows of several species of animals: horses, cows, pigs, etc. He travels to Africa and sees a hippopotamus, but he refuses to accept that this is a new species of animal, maintaining that it is instead an animal "in some ways like a horse and in some ways like a pig". Rather than say "an electron behaves somewhat like a wave and somewhat like a particle", I like to say "an electron behaves exactly like an electron — this behavior is not familiar and you might not be comfortable with it, but that is no reason to denigrate the electron".

> *Technical aside:* When the delayed choice experiments of section 12.2 (page 96) are adapted to atom interferometers, they reinforce the idea that an atom passes through an interferometer not as a classical particle nor as a classical wave, but rather in its own inimitable quantum mechanical fashion.

The Heisenberg uncertainty principle and the phrase "wave-particle duality" are treated with reverence and awe in some circles. But when you get right down to it they really mean nothing more than that an electron is not a small hard marble.

## 14.4   References

This chapter is inspired by the treatment in

> R.P. Feynman, *QED: The Strange Theory of Light and Matter* (Princeton University Press, Princeton, New Jersey, 1985) pages 36–49.

A modern perspective on the Heisenberg uncertainty principle and wave-particle duality is presented by

> Berthold-Georg Englert, Marlan O. Scully, and Herbert Walther, "The duality in matter and light", *Scientific American*, **271** (6) (December 1994) 86–92.

The serendipitous history of the Davisson–Germer experiment is told in

R.K. Gehrenbeck, "Electron diffraction: fifty years ago", *Physics Today*, **31** (1) (January 1978) 34–41. (Be sure to notice also this issue's stunning cover photograph.)

## 14.5   Sample problem

Electrons are shot down toward a crystal of iron. At what speed should they be shot so that a significant number of them are deflected by 90°? (The distance between rows of atoms in iron is $2.87 \times 10^{-8}$ cm; the mass of an electron is $9.11 \times 10^{-28}$ gram.)

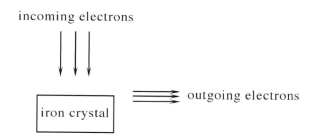

## Solution

As usual, we follow the steps listed on page 78.

*Step 1: Enumerate all the paths from input to output.*

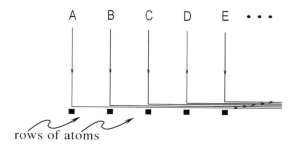

*Step 2: Assign an amplitude arrow to each path.* According to the formula on page 104, the arrow associated with a path rotates this many times:

$1.51 \times 10^{26} \times$ (length in cm) $\times$ (mass in grams) $\times$ (speed in cm/s).

Because each path has a different length, each arrow will rotate a different amount.

*Step 3: Add up all the arrows.* Most circumstances are rather like the one illustrated below. (The amplitude arrow associated with each path is sketched below the letter labeling that path.)

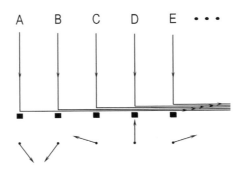

For an electron shot down at this particular speed, the additional length of path A over path B means that the amplitude arrow associated with path A has rotated 70° more than the amplitude arrow associated with path B. The same holds for paths B and C, paths C and D, etc. (I will call this quantity the "excess rotation" of path A over path B, of path B over path C, etc.) Thus the different arrows associated with the many different paths are pointing every-which-way, so when the arrows are added they will mostly cancel out. In such circumstances, the sum arrow will be small and there will be a low probability of deflection by 90°.

Suppose, however, that an electron is shot down faster than the one above was. Then each arrow rotates more than it did above. More importantly, the excess rotation of one path over its shorter neighbor also increases. For a slight increase in speed, there will be a slight increase in excess rotation: say from 70° to 90°. Still, the arrows will be pointing every-which-way and, when added, they will mostly cancel out. But what if there is a significant increase in speed leading to a significant increase in excess rotation, say to 360°, a full rotation?

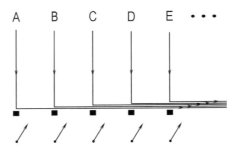

In this case each of the arrows points in *exactly* the same direction, so when they are all added together they produce a large sum arrow and hence a high probability of deflection by 90°.

What is this special speed that results in a large probability of deflection by 90°? It is the speed at which the distance between rows of atoms corresponds to exactly one rotation, that is, the speed at which

$$1 = 1.51 \times 10^{26} \times \text{(distance between rows in cm)} \times \text{(mass in grams)} \times \text{(speed in cm/s.)}.$$

Solving this equation for the speed gives $2.53 \times 10^8$ cm/s. This speed is very large on a human scale, but because electrons have so little mass it is easy to make them go this fast.

## 14.6 Problems

14.1 *Other speeds.* The previous sample problem (section 14.5) finds a speed that gives rise to a substantial probability of deflection by 90° when an electron is shot down at an iron crystal. Will there be a substantial probability of deflection by 90° if an electron of twice this speed is used? Three times? Half the speed? One-third? One-quarter?

14.2 *Curlicue paths.* Consider the motion of a ball from point P to point Q without floors or barriers. Enumerate typical paths between the two points, including curved and three-dimensional paths, and draw representative amplitude arrows associate with each one. Generalize the reasoning on page 105 to show that, in the classical limit, the bundle of paths that are nearly straight lines from P to Q provide most of the amplitude to go from P to Q.

14.3 *Heisenberg uncertainty principle.* In his 1993 Oersted Medal acceptance speech, the distinguished physicist Hans Bethe said

> The [Heisenberg] uncertainty principle simply tells us that the concepts of classical physics are not applicable to the atomic world. But we think in classical terms, and therefore we need the uncertainty principle to reconcile our classical terms with the reality of quantum theory.

Would this passage be improved by replacing the phrase "classical terms" with "classical terminology"? Justify your answer.

14.4 *Wave-particle duality.* On page 57 I summarized the first half of this book by saying that

If $m_z$ has a definite value, then $m_x$ doesn't have a value. If you measure $m_x$, then of course you find some value, but no one (not even the atom itself!) can say with certainty what that value will be — only the probabilities of measuring the various values can be calculated.

Produce a corresponding statement that applies to an electron rather than to the magnetic arrow of a silver atom, and that uses "position" and "speed" rather than "$m_z$" and "$m_x$".

# 15

# The Wavefunction

## 15.1 Between release and detection

In the previous chapter, we talked about finding the probability that a ball released at point P would be detected at point Q. We found out how to calculate this probability by assigning an appropriate amplitude arrow to each of the possible paths from P to Q, and then adding up all the arrows. But, what happens if the ball is released at point P and then detected at some other point, say R? (See the figure below.) You know the procedure for finding this probability: enumerate paths from P to R, assign to each path an amplitude arrow using the formula on page 104, and add up all the arrows. It is somewhat more difficult to execute this procedure for the P to R case than it was for the P to Q case, because it lacks the symmetry. Nevertheless it is clear that many of the same features will apply to both processes: for example, in both cases the largest contribution to the sum amplitude arrow comes from a bundle of paths near the path of minimum length. You might find this problem technically difficult, but it is conceptually straightforward and you could do it if you had to.

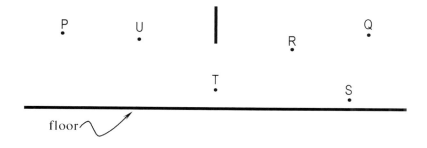

But we don't have to stop here. We could consider having one detector

113

at Q and another detector at R at the same time. Indeed, we could sprinkle detectors all over the page, at points S, T, U, etc. You know how to find an amplitude arrow for the motion from point P to any of these points, and from the arrow you know how to find the transition probability. The figure below shows what these amplitude arrows might look like.

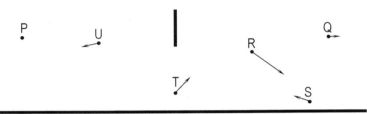

Now, what if the ball is released at point P, we wait four seconds, and *none* of our detectors go off? How are we to describe the state of the ball after it has been released but not yet detected? We can't say "It's at point R" or "It's at point T" because we don't, indeed we *can't*, know what its position is — the ball doesn't have a position. There is only one way to specify the quantal state of the ball between release and detection, and that is by listing the amplitude arrows for all the points where the ball has some amplitude for being, just as in the figure above. This list is called "the wavefunction".

> *Technical aside:* A word concerning etymology is in order here. In mathematics, the word "function" means a set of numbers assigned to every point in space, or to every instant of time, or both. For example, if there are waves on the surface of a pond, then the height of water in the pond is a function of both position and time. As we saw in chapter 8, "Optical interference", a set of arrows very much like amplitude arrows can be related to waves like those on a pond. In the early days of quantum mechanics, this analogy was believed to be much stronger than it actually is, so the list of amplitude arrows was named the "wave function". In recognition of the important differences that we now recognize between classical waves and amplitude arrows, today the two words are usually closed up as "wavefunction".

Above we supposed that the ball was released from point P and not detected for four seconds. What would happen if, at the five-second mark, the ball were detected at point T? How do we describe the ball's state the instant after it is detected? The answer is simple: we just say that it is located at point T. We no longer need to keep track of the amplitude arrows at points Q, R, S, etc., because although the ball could have gone to any of them, it didn't. Thus immediately before detection the state of the ball is specified by a bunch of arrows spread over many points, while immediately after detection it is specified by giving just one point. What happened to all those arrows? Nothing happened to them, because they never were there. They were never anything more than mathematical tools to help keep our calculations straight. The process described above is called "the collapse of the wavefunction", and it greatly worries those who think that the amplitude arrows are somehow physically out in space, in the same way that air molecules are physically out in space. You don't have that misimpression, so the collapse shouldn't bother you at all.

## 15.2    What does an electron look like?

The literal answer to this question is "It doesn't look like anything. An electron is too small to be seen." This answer is in fact the dominant one found in discussions on quantum mechanics. We are told not to ask questions that cannot be answered through direct experiment.* For example if an electron is released at point P and detected at point Q, and moves between the points in total darkness so that it is not possible, even in principle, to determine which route it took in moving from P to Q, then we are told that is is not proper to ask which route it took.

This dominant answer is correct but, in my experience, unsatisfactory. When we ask "What does an electron look like?" we really mean "What is the character (or nature) of an electron?" or "How does an electron behave?" or "How can an electron be visualized?". Humans are visual animals, and even if we are told not to visualize a phenomenon we do so anyway — the pictures just pop into our minds unbidden. In quantum mechanics this often leads to naive and incorrect visualizations, which people continue to carry in their minds precisely because the dominant position encourages them not to critically examine their visualizations. So rather than just ignore the issue I like to face it head on, acknowledging that our classical minds are unlikely to produce perfectly accurate visualizations, but realizing that an imperfect visualization, with its imperfections understood, is far superior to an imperfect visualization which is held uncritically. To paraphrase Socrates, "the unexamined visualization is not worth visualizing".

Let us return to the electron moving from P to Q in complete darkness.†

---

* For example: "The single electron *does* interfere with itself. But don't try to visualize how it does so!"

† In technical terms, this paragraph and the next point out the difficulty of visualizing the quantal wavefunction in view of the facts that (i) the wavefunction is complex valued and (ii) it exists in

At the instants of its release and its detection the electron behaves like a very small, very hard marble, in that it has a definite position. But between these two events the electron doesn't have a definite position. Sometimes I visualize it as a cloud that is thicker at places where the electron is more likely to be and thinner at places where it is less likely to be. This visualization captures beautifully the probabilistic character of quantum mechanics, but it shows nothing of the interference character. So I sometimes visualize an electron instead through a swarm of rotating amplitude arrows, the swarm being thicker and the arrows longer where the electron is more likely to be. This can give me nightmares, so more often I simply modify the cloud visualization by assigning colors to different arrow directions and mentally coloring each point of the cloud according to the direction of the amplitude arrow there. In my mind's eye, I see the electron as a swirl of shimmering colors. Both of these visualizations can be useful, but both have the defect of infusing a mathematical tool — the amplitude arrow — with physical reality.

The problem becomes even more acute when one attempts to visualize a system of two particles because then (see section 11.2) one must visualize not one state for one particle and another state for the other particle, but instead a single state for the pair of particles.

It is easier to show why some visualizations are poor than to produce visualizations that are good. For example, some people like to visualize an electron as a small hard marble that takes a definite and well-defined route from P to Q, but that the actual route to be taken is not predictable beforehand, so that sometimes the marble will take one route and sometimes it will take another. It is impossible, however, to make such a picture consistent with the interference results of chapter 9. (Or at least, it is impossible to do so without invoking mysterious messages that allow a marble passing through branch a to know whether or not branch b is open or blocked.) So you may not know what an electron looks like, but at least you know what it doesn't look like!

The problem of visualization is closely connected to a problem of terminology. To many, the word "particle" conjures up the image of a small, hard, classical marble. In quantum mechanics, it is not entirely clear what the image associated with "particle" ought to be, but it most certainly is *not* this classical picture! If we were eminent Victorians we would find a noble Latin root and build a new word to describe the quantal particle. If we lived in Washington, DC, or Arlington, Virginia (the site of the Pentagon), we would invent an acronym (something like PAWBITQMF — particle and/or wave behaving in typical quantum mechanical fashion).

---

configuration space *or* in momentum space but *not* in ordinary three-dimensional position space.

There have been some attempts to coin a new word: "wavicle", "quon", or "quanton". These attempts have not caught on.

In conclusion, I do not have a visualization — or even a name — that is satisfactory for even so simple a thing as a single quantal electron. Because my mind is filled with classical images and intuition, this is perhaps not surprising. A truly successful visualization would be very close to a classical "clockwork" mechanism that underlies quantum mechanics, and we have already seen (section 9.8) that such a mechanism does not exist. But this lack of visualization must be regarded as a limitation of my imagination, and not as any defect in nature or in quantum mechanics.

*Our imagination is stretched to the utmost, not, as in fiction, to imagine things which are not really there, but just to comprehend those things which are there.*

— Richard Feynman

## 15.3   Problems

15.1  *Mistaken visualization.* What is wrong with the statement "Between release and detection, the electron might be at any one of many points"? Can you rephrase the statement to make it correct?

15.2  *Wording.* On 28 May 1996 the New York *Times* published an article titled "Team of physicists proves atom can exist in two places at once". The article describes an experiment in which Chris Monroe and coworkers "succeeded in separating two states of a single atom in space, then pulled them 83 nanometers apart". This article's title is perfectly appropriate for an audience unfamiliar with quantum mechanics and its terminology. Now that you do know the terminology of quantum mechanics, think up a more accurate title.

15.3  *Visualization.* On page 176 of his book *In Search of Schrödinger's Cat*, John Gribbin claims that electron interference raises "the puzzle that an electron at hole A *knows* whether hole B is open or closed". Which incorrect visualization of an electron is Gribbin using that makes this phenomenon seem puzzling to him?

15.4  *Need for visualization.* Does our inability to find a satisfactory visualization for a quantal particle mean that the dominant position ("don't ask questions that you can't answer") is the best one after all? Is its absence merely distressing or does it constitute a fundamental flaw in our knowledge? (Let me point out that distressing things are, by definition, not pleasant, but neither, unfortunately, are they rare.)

15.5 *Measurement.* Mr. Parker finds the quantum measurement process difficult to understand. "Suppose I start with an atom in a state so that it has equal probability of being anywhere in a box. If I shine a strong light throughout the entire box I will find the atom only at one point. But what happens if I shine the light on only the left half of the box, and don't find the atom? I now know that the atom is somewhere in the right half. How could the light, shining where the atom isn't, affect the atom?" Convince Mr. Parker that the conflict is not between quantum mechanics and reality, but between quantum mechanics and his incorrect visualization of the atom as a tiny marble. (This conundrum is called the Renninger negative-result experiment.)

15.6 *Visualization techniques.* (For technical readers.) This chapter mentioned two techniques for visualizing wavefunctions: through a swarm of amplitude arrows ("phasors") and through color. I have written a computer program that displays one-dimensional time-varying wavefunctions using either of these techniques, and two other techniques as well. Download the program (it works under the MS-DOS operating system) through the World Wide Web site mentioned on page xiv, and evaluate these different display styles. Can you come up with new visualization techniques of your own? If so, please tell me what they are!

15.7 *Faster-than-light propagation.* (For technical readers.) In a one-particle situation in quantum mechanics, the wavefunction at a given point changes instantly as soon as the particle is detected. In the Coulomb gauge, the electric potential (and the vector potential) at a given point changes the instant that any charged particle, anywhere in the universe, is moved. Does either mechanism permit instantaneous communication?

# Appendix A

## A Brief History of Quantum Mechanics

Up to now this book has focused on the behavior of nature. I could say more: more about measurement, more about the classical limit, more about different rules for assigning amplitudes, and so forth, but the main points have been made. So instead of talking more about nature I'm going to talk about people — about how people discovered quantum mechanics.

### A.1 Warnings

I am not a historian of science. The history of science is a very difficult field. A historian of science must be just as proficient at science as a scientist is, but must also have a good understanding of personalities, and a good knowledge of the social and political background that affects developments in science and that is in turn affected by those developments. He or she has to know not only the outcome of the historical process, namely the science that is generally accepted today, but also the many false turns and blind alleys that scientists tripped across in the process of discovering what we believe today. He or she must understand not only the cleanest and most direct experimental evidence supporting our current theories (like the evidence presented in this book), but must understand also how those theories came to be accepted through a tightly interconnected web of many experiments, no one of which was completely convincing but which taken together presented an overwhelming argument.

Thus a full history of quantum mechanics would have to discuss Schrö-dinger's many mistresses, Ehrenfest's suicide, and Heisenberg's involvement with Nazism. It would have to treat the First World War's effect on the development of science. It would need to mention "the Thomson model" of the atom, which was once the major competing theory to quantum mechanics. It would have to give appropriate weight to both theoretical and experimental developments. Needless to say, such a complete history will never be written, and this brief appendix will not even broach most of these topics. The references on page 131 will lead you to further information.

The historian of science has problems beyond even these. The work of government is generally carried out through the exchange of written memos, and when verbal arguments are used (as in Congressional hearings) detailed written transcripts are maintained. These records are stored in archives to insure that historians interested in government decisions will have access to them. Historians of science do not have such advantages. Much of the work of science is done through informal conversations, and the resulting written record is often sanitized to avoid offending competing scientists. The invaluable oral record is passed down from professor to student repeatedly before anyone ever records it on paper. Naturally, the stories tend to become better and better as they are transmitted over and over. In addition, there is a tendency for the exciting stories to be repeated and the dull ones to be forgotten, leading to a Darwinian "survival of the funniest" — rather than of the most accurate.

Finally, once all the historical records have been sifted and analyzed, there remains the problem of overall synthesis and presentation. Many scientific historians (and even more scientists) like to tell a story in which each step follows naturally from the one preceding it, scientists always work cooperatively and selflessly, and where harmony rules.* Such stories infuriate me. They remind me of the stock market analysts who come onto television every evening and explain in detail the cause of every dip and curve in the Dow for the preceding day. If they know the stock market so well, why do they wait until evening to tell me about it? Why don't they tell me in the morning so that it can do me some good? For that matter, why are they on television at all, rather than out relaxing on their million-dollar yachts? The fact is that scientific history, like the stock market and like everyday life, does not proceed in an orderly, coherent pattern. The story of quantum mechanics is a story full of serendipity, personal squabbles, opportunities missed and taken, and of luck both good and bad.

Because I find the sugar-sweet stories of the harmonious development of science to be so offensive, when I tell the story I emphasize the conflicts, the contingencies, and the unpredictablities. Hence the story I tell is no more accurate than the sweet talk, because I go too far in the opposite direction. Keep in mind, as you read the story that follows, that I suffer

---

* I told a story like this myself in section 10.2, "Evidence for the amplitude framework", where I suggested that discoveries in physics always result from the exploration of shorter length scales. In fact, discoveries also come from the exploration of longer length scales, of lower temperatures, of greater complexity, and simply by investigating familiar phenomena in more detail.

from this overreaction as well as all the other difficulties mentioned in this section.

## A.2    Status of physics: January 1900

In January 1900 the atomic hypothesis was widely but not universally accepted. Atoms were considered point particles, and it wasn't clear how atoms of different elements differed. The electron had just been discovered (1897) and it wasn't clear where (or even whether) electrons were located within atoms. One important outstanding problem concerned the colors emitted by atoms in a discharge tube (familiar today as the light from a fluorescent tube or from a neon sign). No one could understand why different gas atoms glowed in different colors. Another outstanding problem concerned the amount of heat required to change the temperature of a diatomic gas such as oxygen: the measured amounts were well below the value predicted by theory. Because quantum mechanics is important when applied to atomic phenomena, you might guess that investigations into questions like these would give rise to the discovery of quantum mechanics. Instead it came from a study of heat radiation.

## A.3    Heat radiation

You know that the coals of a campfire, or the coils of an electric stove, glow red. You probably don't know that even hotter objects glow white, but this fact is well known to blacksmiths. When objects are hotter still they glow blue. (This is why a gas stove should be adjusted to make a blue flame.) Indeed, objects at room temperature also glow (radiate), but the radiation they emit is infrared, which is not detectable by the eye. (The military has developed — for use in night warfare — special eye sets that convert infrared radiation to optical radiation.)

These observations can be explained qualitatively by thinking of heat as a jiggling of atoms: like jello, but on a smaller scale so that you can't see the vibrations due to heat. At higher temperatures the atoms jiggle both farther and faster. The increased distance of jiggling accounts for the brighter radiation from hotter bodies, while the increased speed accounts for the change in color.

In the year 1900 several scientists were trying to turn these observations into a detailed explanation of and a quantitatively accurate formula for the color of heat radiation as a function of temperature. On 19 October 1900 the Berliner Max Planck (age 42) announced a formula that fit the experimental results perfectly, yet he had no explanation for the formula — it just happened to fit. He worked to find an explanation through the late fall and finally was able to derive his formula by assuming that the atomic jigglers could not take on any possible energy, but only certain special "allowed" values. He announced this result on 14 December 1900. This date is now considered the birthday of quantum mechanics (and there is certain to be a big celebration on its one hundredth anniversary) but at the time no one found it particularly significant. We know this not only from contemporary reports, but also because the assumption of allowed energy values raises certain obvious questions that no one bothered to follow up. For example, how does the jiggler change from one allowed energy to another if the intermediate energies are prohibited? Again, if a jiggling atom can only assume certain allowed values of energy, then there must also be restrictions on the positions and speeds that the atom can have. What are they? Planck never tried to find out.

Thirty-one years after his discovery Planck wrote:

> I can characterize the whole procedure as an act of desperation, since, by nature I am peaceable and opposed to doubtful adventures. However, I had already fought for six years (since 1894) with the problem of equilibrium between radiation and matter without arriving at any successful result. I was aware that this problem was of fundamental importance in physics, and I knew the formula describing the energy distribution ... hence a theoretical interpretation *had* to be found at any price, however high it might be.

It should be clear from what I have already said that this is just a beautiful and romantic story that was developed with good thirty-year hindsight. Here is another wonderful story, this one related by Werner Heisenberg:

> In a period of most intensive work during the summer of 1900 [Planck] finally convinced himself that there was no way of escaping from this conclusion [of "allowed" energies]. It was told by Planck's son that his father spoke to him about his new ideas on a long walk through the Grunewald, the wood in the suburbs of Berlin. On this walk he explained that he felt he had possibly made a discovery of the first rank, comparable perhaps only to the discoveries of Newton.

As much as I would like for this beautiful story to be true, the intensive work took place during the late fall, not the summer, of 1900. If Planck did indeed take his son for a long walk on the afternoon that he discovered quantum mechanics, the son would probably remember the nasty cold he caught better than any remarks his father made.

## A.4    The old quantum theory

Although the ideas of Planck did not take the world by storm, they did develop a growing following and were applied to more and more situations. The resulting ideas, now called "old quantum theory", were all of the same type: Classical mechanics was assumed to hold, but with the additional assumption that only certain values of a physical quantity (the energy, say, or the projection of a magnetic arrow) were allowed. Any such quantity was said to be "quantized". The trick seemed to be to guess the right quantization rules for the situation under study, or to find a general set of quantization rules that would work for all situations.

For example, in 1905 Albert Einstein (age 26) postulated that the total energy of a beam of light is quantized. Just one year later he used quantization ideas to explain the heat/temperature puzzle for diatomic gases. Five years after that, in 1911, Arnold Sommerfeld (age 43) at Munich began working on the implications of energy quantization for position and speed.

In the same year Ernest Rutherford (age 40), a New Zealander doing experiments in Manchester, England, discovered the atomic nucleus — only at this relatively late stage in the development of quantum mechanics did physicists have even a qualitatively correct picture of the atom! In 1913, Niels Bohr (age 28), a Dane who had recently worked in Rutherford's laboratory, introduced quantization ideas for the hydrogen atom. His theory was remarkably successful in explaining the colors emitted by hydrogen glowing in a discharge tube, and it sparked enormous interest in developing and extending the old quantum theory.

This development was hindered but not halted completely by the start of the First World War in 1914. During the war (in 1915) William Wilson (age 40, a native of Cumberland, England, working at King's College in London) made progress on the implications of energy quantization for position and speed, and Sommerfeld also continued his work in that direction.

With the coming of the armistice in 1918, work in quantum mechanics expanded rapidly. Many theories were suggested and many experiments performed. To cite just one example, in 1922 Otto Stern and his graduate student Walther Gerlach (ages 34 and 23) performed their important experiment that is so essential to the way this book presents quantum mechanics. Jagdish Mehra and Helmut Rechenberg, in their monumental history of quantum mechanics, describe the situation at this juncture well:

> At the turn of the year from 1922 to 1923, the physicists looked forward with enormous enthusiasm towards detailed solutions of the outstanding problems, such as the helium problem and

the problem of the anomalous Zeeman effects. However, within less than a year, the investigation of these problems revealed an almost complete failure of Bohr's atomic theory.

## A.5   The matrix formulation of quantum mechanics

As more and more situations were encountered, more and more recipes for allowed values were required. This development took place mostly at Niels Bohr's Institute for Theoretical Physics in Copenhagen, and at the University of Göttingen in northern Germany. The most important actors at Göttingen were Max Born (age 43, an established professor) and Werner Heisenberg (age 23, a freshly minted Ph.D. from Sommerfeld in Munich). According to Born "At Göttingen we also took part in the attempts to distill the unknown mechanics of the atom out of the experimental results. ... The art of guessing correct formulas ... was brought to considerable perfection."

Heisenberg particularly was interested in general methods for making guesses. He began to develop systematic tables of allowed physical quantities, be they energies, or positions, or speeds. Born looked at these tables and saw that they could be interpreted as mathematical matrices. Fifty years later matrix mathematics would be taught even in high schools. But in 1925 it was an advanced and abstract technique, and Heisenberg struggled with it. His work was cut short in June 1925. As Mehra and Rechenberg describe it:

> This was late spring in Göttingen, with fresh grass and flowering bushes, and Heisenberg was interrupted in his work by a severe attack of hay fever. Since he could hardly do anything, he had to ask his director, Max Born, for a leave of about two weeks, which he decided to spend on the rocky island of Helgoland to effect a cure.
>
> On 7 June 1925 Heisenberg took the night train from Göttingen to Cuxhaven where he had to catch the ferryboat for Helgoland in the morning. On arrival at Cuxhaven, "I was extremely tired and my face was swollen. I went to get breakfast in a small inn and the landlady said, 'You must have had a pretty bad night. Somebody must have beaten you.' She thought I had had a fight with somebody. I told her that I was ill and that I had to take the boat, but she was still worried about me." A few hours later he reached Helgoland.
>
> Helgoland, a rocky island in the North Sea, consists of a mass of red sandstone, rising abruptly to an elevation of about 160 feet, and there is nearly no vegetation on it. [It has an

area of about 380 acres and a permanent population of several hundred inhabitants. On the lower section of the island lies a fishing village, while the upper section serves as a summer resort for tourists. ... From 1402 to 1714 it formed a part of Schleswig-Holstein, then became Danish until it was seized by the English fleet in 1807. It was formally ceded to Great Britain in 1814. Britain gave it to Germany in exchange for Zanzibar and some territory in Africa (1890). Helgoland was an important base for the German Navy. In accordance with the Treaty of Versailles the military and naval fortifications were demolished in 1920–1922. Under the Nazi régime Helgoland again became a military stronghold and was a target for heavy Allied bombing towards the end of World War II. From 1947 to 1 March 1952, when it was handed back to Germany, the island was used as a bombing range by the Royal Air Force. Then it was restored as a tourist and fishing center.] Heisenberg rented a room on the second floor of a house situated high above the southern edge of the island, which offered him a "glorious view over the village, and the dunes and the sea beyond." "As I sat on my balcony," he recalled more than forty years later, "I had ample opportunity to reflect on Bohr's remark that part of infinity seems to lie within the grasp of those who look across the sea." He began to take walks to the upper end of the island and swam daily in the sea. Soon he felt much better, and he began to divide his time into three parts. The first he still used for walking and swimming; the second he spent in reading Goethe's *West-Östlicher Divan*; and the third he devoted to work on physics. Having nothing else to distract him, he could reflect with great concentration on the problems and difficulties which had been occupying him until a few days earlier in Göttingen.

Heisenberg reproduced his earlier work, cleaning up the mathematics and simplifying the formulation. He worried that the mathematical scheme he invented might prove to be inconsistent, and in particular that it might violate the principle of the conservation of energy. In Heisenberg's own words:

One evening I reached the point where I was ready to determine the individual terms in the energy table, or, as we put it today, in the energy matrix, by what would now be considered an extremely clumsy series of calculations. When the first terms seemed to accord with the energy principle, I became rather excited, and I began to make countless arithmetical errors. As a result, it was almost three o'clock in the morning before the final result of my computations lay before me. The energy principle had held for all the terms, and I could no longer doubt the mathematical consistency and coherence of the kind of quantum mechanics to which my calculations pointed. At first, I was deeply alarmed. I had the feeling that, through the surface of atomic phenomena, I was looking at a strangely beautiful interior, and felt almost giddy at the thought that I now had to probe this wealth of mathematical structures nature had so generously spread out before me. I was far too excited to sleep, and so, as a new day dawned, I made for the southern tip of the island, where I had been longing to climb a rock jutting out into the sea. I now did so without too much trouble, and waited for the sun to rise.

By the end of the summer Heisenberg, Born, and Pascual Jordan (age 22) had developed a complete and consistent theory of quantum mechanics. (Jordan had entered the collaboration when he overheard Born discussing quantum mechanics with a colleague on a train.)

This theory, called "matrix mechanics" or "the matrix formulation of quantum mechanics", is not the theory I have presented in this book. It is extremely and intrinsically mathematical, and even for master mathematicians it was difficult to work with. Although we now know it to be complete and consistent, this wasn't clear until much later. Heisenberg had been keeping Wolfgang Pauli apprised of his progress. (Pauli, age 25, was Heisenberg's friend from graduate student days, when they studied together under Sommerfeld.) Pauli found the work too mathematical for his tastes, and called it "Göttingen's deluge of formal learning". On 12 October 1925 Heisenberg could stand Pauli's biting criticism no longer. He wrote to Pauli:

> With respect to both of your last letters I must preach you a sermon, and beg your pardon for proceeding in Bavarian: It is really a pigsty that you cannot stop indulging in a slanging match. Your eternal reviling of Copenhagen and Göttingen is a shrieking scandal. You will have to allow that, in any case, we are not seeking to ruin physics out of malicious intent. When you reproach us that we are such big donkeys that we have never produced anything new in physics, it may well be true. But then, you are also an equally big jackass because you have not accomplished it either......(The dots denote a curse of about two-minute duration!) Do not think badly of me and many greetings.

## A.6   The wavefunction formulation of quantum mechanics

While this work was going on at Göttingen and Helgoland, others were busy as well. In 1923 Louis de Broglie (age 31), associated an "internal periodic phenomenon" — a wave — with a particle. He was never very precise about just what that meant. (De Broglie is sometimes called "Prince de Broglie" because his family descended from the French nobility. To be strictly correct, however, only his eldest brother could claim the title.)

It fell to Erwin Schrödinger, an Austrian working in Zürich, to build this vague idea into a theory of wave mechanics. He did so during the Christmas season of 1925 (at age 38), at the alpine resort of Arosa, Switzerland, in the company of "an old girlfriend [from] Vienna", while his wife stayed home in Zürich.

In short, just twenty-five years after Planck glimpsed the first sight of a new physics, there was not one, but two competing versions of that new physics! The two versions seemed utterly different and there was an acrimonious debate over which one was correct. In a footnote to a 1926 paper Schrödinger claimed to be "discouraged, if not repelled" by matrix mechanics. Meanwhile, Heisenberg wrote to Pauli (8 June 1926) that

> The more I think of the physical part of the Schrödinger theory, the more detestable I find it. What Schrödinger writes about visualization makes scarcely any sense, in other words I think it is shit. The greatest result of his theory is the calculation of matrix elements.

Fortunately the debate was soon stilled: in 1926 Schrödinger and, independently, Carl Eckert (age 24) of Caltech proved that the two new mechanics, although very different in superficial appearance, were equivalent to each other.[†] (Pauli also proved this, but never published the result.)

## A.7   Applications

With not just one, but two complete formulations of quantum mechanics in hand, the quantum theory grew explosively. It was applied to atoms, molecules, and solids. It solved with ease the problem of helium (see page 123) that had defeated the old quantum theory. It resolved questions concerning the structure of stars, the nature of superconductors, and the properties of magnets. One particularly important contributor was

---

[†] Very much as the process of adding arabic numerals is very different from the process of adding roman numerals, but the two processes nevertheless always give the same result (see problem 8.2).

P.A.M. Dirac, who in 1926 (at age 22) extended the theory to relativistic and field-theoretic situations. Another was Linus Pauling, who in 1931 (at age 30) developed quantum mechanical ideas to explain chemical bonding, which previously had been understood only on empirical grounds. Even today quantum mechanics is being applied to new problems and new situations. It would be impossible to mention all of them. All I can say is that quantum mechanics, strange though it may be, has been tremendously successful.

### A.8    The Bohr–Einstein debate

The extraordinary success of quantum mechanics in applications did not overwhelm everyone. A number of scientists, including Schrödinger, de Broglie, and — most prominently — Einstein, remained unhappy with the standard probabilistic interpretation of quantum mechanics. In a letter to Max Born (4 December 1926), Einstein made his famous statement that

> Quantum mechanics is very impressive. But an inner voice tells me that it is not yet the real thing. The theory produces a good deal but hardly brings us closer to the secret of the Old One. I am at all events convinced that *He* does not play dice.

In concrete terms, Einstein's "inner voice" led him, until his death, to issue occasional detailed critiques of quantum mechanics and its probabilistic interpretation. Niels Bohr undertook to reply to these critiques, and the resulting exchange is now called the "Bohr–Einstein debate". At one memorable stage of the debate (Fifth Solvay Congress, 1927), Einstein made an objection similar to the one quoted above and Bohr

> replied by pointing out the great caution, already called for by ancient thinkers, in ascribing attributes to Providence in every-day language.

These two statements are often paraphrased as, Einstein to Bohr: "God does not play dice with the universe." Bohr to Einstein: "Stop telling God how to behave!" While the actual exchange was not quite so dramatic and quick as the paraphrase would have it, there was nevertheless a wonderful rejoinder from what must have been a severely exasperated Bohr.

The Bohr–Einstein debate had the benefit of forcing the creators of quantum mechanics to sharpen their reasoning and face the consequences of their theory in its most starkly non-intuitive situations. It also had (in my opinion) one disastrous consequence: because Einstein phrased his objections in purely classical terms, Bohr was compelled to reply in nearly classical terms, giving the impression that in quantum mechanics, an electron is "really classical" but that somehow nature puts limits on how well we can determine those classical properties. I have tried in this book to convince you that this is a misconception: the reason we cannot measure simultaneously the exact position and speed of an electron is because an electron does *not have* simultaneously an exact position and speed. It is no defect in our measuring instruments that they cannot measure what does not exist. This is simply the character of an electron — an electron is *not* just a smaller, harder edition of a marble. This misconception — this picture of a classical world underlying the quantum world — poisoned my own understanding of quantum mechanics for years. I hope that you will be able to avoid it.

On the other hand, the Bohr–Einstein debate also had at least one salutary product. In 1935 Einstein, in collaboration with Boris Podolsky and Nathan Rosen, invented a situation in which the results of quantum mechanics seemed completely at odds with common sense, a situation in which the measurement of a particle at one location could reveal instantly information about a second particle far away. The three scientists published a paper which claimed that "No reasonable definition of reality could be expected to permit this." Bohr produced a recondite response and the issue was forgotten by most physicists, who were justifiably busy with the applications of rather than the foundations of quantum mechanics. But the ideas did not vanish entirely, and they eventually raised the interest of John Bell. In 1964 Bell used the Einstein–Podolsky–Rosen situation to produce a theorem about the results from certain distant measurements for any deterministic scheme, not just classical mechanics. In 1982 Alain Aspect and his collaborators put Bell's theorem to the test and found that nature did indeed behave in the manner that Einstein (and others!) found so counterintuitive.

## A.9   The amplitude formulation of quantum mechanics

The version of quantum mechanics presented in this book is neither matrix nor wave mechanics. It is yet another formulation, different in approach and outlook, but fundamentally equivalent to the two formulations already mentioned. It is called amplitude mechanics (or "the sum over histories technique", or "the many paths approach", or "the path integral formulation", or "the Lagrangian approach", or "the method of least action"), and it was developed by Richard Feynman in 1941 while he was a graduate student (age 23) at Princeton. Its discovery is well described by Feynman himself in his Nobel lecture:

> I went to a beer party in the Nassau Tavern in Princeton.
> There was a gentleman, newly arrived from Europe (Herbert

Jehle[‡]) who came and sat next to me. Europeans are much more serious than we are in America because they think a good place to discuss intellectual matters is a beer party. So he sat by me and asked, "What are you doing" and so on, and I said, "I'm drinking beer." Then I realized that he wanted to know what work I was doing and I told him I was struggling with this problem, and I simply turned to him and said "Listen, do you know any way of doing quantum mechanics starting with action — where the action integral comes into the quantum mechanics?" "No," he said, "but Dirac has a paper in which the Lagrangian, at least, comes into quantum mechanics. I will show it to you tomorrow."

Next day we went to the Princeton Library (they have little rooms on the side to discuss things) and he showed me this paper.

Dirac's short paper in the *Physikalische Zeitschrift der Sowjetunion* claimed that a mathematical tool which governs the time development of a quantal system was "analogous" to the classical Lagrangian.

Professor Jehle showed me this; I read it; he explained it to me, and I said, "What does he mean, they are analogous; what does that mean, *analogous?* What is the use of that?" He said, "You Americans! You always want to find a use for everything!" I said that I thought that Dirac must mean that they were equal. "No," he explained, "he doesn't mean they are equal." "Well," I said, "let's see what happens if we make them equal."

So, I simply put them equal, taking the simplest example ... but soon found that I had to put a constant of proportionality $A$ in, suitably adjusted. When I substituted ... and just calculated things out by Taylor-series expansion, out came the Schrödinger equation. So I turned to Professor Jehle, not really understanding, and said, "Well you see Professor Dirac meant that they were proportional." Professor Jehle's eyes were bugging out — he had taken out a little notebook and was rapidly copying it down from the blackboard and said, "No, no, this is an important discovery."

Feynman's thesis advisor, John Archibald Wheeler (age 30), was equally impressed. He believed that the amplitude formulation of quantum me-

---

[‡] Jehle had been a student of Schrödinger in Berlin, and was in Princeton fleeing the Nazis. He was a Quaker and had survived prison camps in both Germany and France.

chanics — although mathematically equivalent to the matrix and wave formulations — was so much more natural than the previous formulations that it had a chance of convincing quantum mechanics's most determined critic. Wheeler writes:

> Visiting Einstein one day, I could not resist telling him about Feynman's new way to express quantum theory. "Feynman has found a beautiful picture to understand the probability amplitude for a dynamical system to go from one specified configuration at one time to another specified configuration at a later time. He treats on a footing of absolute equality every conceivable history that leads from the initial state to the final one, no matter how crazy the motion in between. The contributions of these histories differ not at all in amplitude, only in phase. ... This prescription reproduces all of standard quantum theory. How could one ever want a simpler way to see what quantum theory is all about! Doesn't this marvelous discovery make you willing to accept the quantum theory, Professor Einstein?" He replied in a serious voice, "I still cannot believe that God plays dice. But maybe", he smiled, "I have earned the right to make my mistakes."

## A.10    References

Banesh Hoffmann, *The Strange Story of the Quantum* (Dover, New York, 1959). A popular history.

Barbara L. Cline, *Men Who Made a New Physics* (University of Chicago Press, Chicago, Illinois, 1987). Another popular history, emphasizing biography.

George Gamow, *Thirty Years that Shook Physics* (Doubleday, New York, 1966). Naive and uncritical as history, but wonderful as storytelling.

Abraham Pais, *Inward Bound: Of Matter and Forces in the Physical World* (Clarendon Press, Oxford, UK, 1986). A general history of twentieth century atomic, nuclear, and elementary particle physics. Good on experiment. See particularly chapter 9(c) on alternatives to the nuclear model of the atom. Unfortunately dominated, as the title indicates, by the misconception that the only advances in physics worth mentioning are those opened up by exploring smaller length scales.

Stephen G. Brush, *Statistical Physics and the Atomic Theory of Matter from Boyle and Newton to Landau and Onsager* (Princeton University Press, Princeton, New Jersey, 1983). A more technical history of quantum mechanics. See particularly chapters 4 and 5 for the diverse range of applications of quantum mechanics.

Max Jammer, *The Conceptual Development of Quantum Mechanics*, second edition (American Institute of Physics, New York, 1989). The most comprehensive single-volume history of quantum mechanics.

Jagdish Mehra and Helmut Rechenberg, *The Historical Development of Quantum Theory* (Springer–Verlag, New York, 1982–1987). A very complete history concerning both general and technical points. Seven volumes are already printed and more are on their way.

# Appendix B
## Putting Weirdness to Work

According to Charles de Gaulle, Napoleon's military genius lay in his ability "to grasp the situation, to adapt himself to it, and to exploit it to his own advantage". Most of this book has treated the first two of these steps: learning what quantum mechanics is and how to work with it, whether we like it or not. This appendix moves on to the third step of exploitation.

The applications of quantum mechanics are myriad. Quantum mechanics underlies all chemical and biochemical reactions, the design of drugs and of alloys, and the generation of medical X-rays. It is essential to the laser, to the transistor, and to a sensitive detector of magnetic field called the SQUID (Superconducting QUantum Interference Device). But for the purposes of this book, it is useful to focus on only three of these applications: quantum cryptography, tunneling applications, and quantum computers. The first of these was treated in chapter 13; this appendix describes the second and third. These descriptions are segregated into an appendix because I don't know how to treat them thoroughly at the mathematical level of this book. Consequently, the treatments here are more descriptive and less analytic than the treatments in the chapters.

### B.1 Tunneling

A classical ball rolls in a bowl. Can the ball escape? As the ball rolls up the side of the bowl, it slows down. If the ball has enough energy, it will slow down but not stop, and hence can make it over the side and out. A ball with a low enough energy will always remain inside the bowl.

Is there any difference if we use a quantal ball? In this case, as we have seen, the ball might not have a definite position, so there are situations in which it has some amplitude for being inside the bowl and some amplitude for being outside the bowl. It is also true (although we have not demonstrated this) that the ball might not have a definite energy, so there are situations in which the average energy is too small for the ball to escape, but yet there is some amplitude for the ball to have enough energy to escape. Thus it can happen that a quantal ball starts well inside the bowl with an average energy too small for classical escape, yet nevertheless the ball escapes. This process is called tunneling, because it is a way to get out of a barrier without going over the barrier. (The name unfortunately suggests that the quantal ball bores a hole through the side of the bowl. It doesn't — the bowl is unaltered.)

Are there any practical applications for tunneling? Prisoners might hope to tunnel through the walls of their jail cells, but this is not a practical application: the probability of tunneling through a barrier decreases dramatically with the thickness of the barrier. But this same feature that makes tunneling impractical for prison escape is essential for a device that locates atoms. In this device a thin needle moves across the surface of a sample. Electrons can tunnel from the needle to the sample, but only if the two are very close. In this way, a very precise picture of the sample's surface can be build up. This device, called a "scanning tunneling microscope", can easily locate individual atoms.

Tunneling is also important in the decay of atomic nuclei, for an esoteric electronic component called the "tunnel diode", and as a possible mechanism for superconductivity at high temperatures. My favorite application of tunneling, however, is far from recondite.

The sun produces light energy through a series of nuclear reactions. The first step in this series is that two protons come very close to each other and react to form a proton and neutron bound together, plus a positron, plus a neutrino. If you don't know what a positron is, don't worry. The important thing is that the two protons have to come close together. But the two protons have the same electric charge, so they repel each other strongly. Calculations based on classical mechanics predict that this reaction would happen so slowly that almost no light would come from the sun. A correct calculation based on quantum mechanics shows that one proton tunnels through the barrier of repulsion separating the two, and allows the reaction to proceed.

Quantum mechanics applies to the domain of the very small, but sometimes small things have big consequences. Sunshine itself is generated through the workings of quantum mechanics.

## B.2   Quantum computers

Not so many years ago, it was customary to interpret the Heisenberg uncertainty principle as a limitation on information: "In classical mechanics one can know a particle's position and its speed exactly, but in quantum mechanics one cannot have this complete information." This is quite the wrong attitude. In fact, one may have complete information concerning either a classical state or a quantal state, but the information is different in the two cases. Consider, for example, a single bead strung on a fixed wire. In classical mechanics, the bead's state is specified by listing its position and its speed: two numbers. In quantum mechanics, the bead's state is specified (see chapter 15, "The wavefunction") by listing the amplitude for it to be at any of the points along the wire. Since there are an infinite number of points on the wire, and since the amplitude at each point is specified through two numbers (a magnitude and an angle), specifying a quantal state actually requires considerably *more* information than does specifying a classical state.

In short, the information needed to specify a quantal state is not only different in character from the information needed to specify a classical state, but it is also much larger in quantity. Thus there are many more quantal states than there are classical states for the same system. This fact is a source of both delight and difficulty. The delight stems from the great richness and variety of quantal behavior, a variety lacking in the classical domain simply because there are many more ways to be quantal than there are ways to be classical. The difficulty lies in the fact that calculations involving quantal systems necessarily process a lot more information than those involving the corresponding classical system, and thus are usually more difficult to perform. A computer program simulating a quantal system will almost always run slower than one simulating the corresponding classical system: the quantum simulation simply has more information to keep track of.

For many years, this was regarded as an unpleasant but unavoidable fact of scientific life. Then, in the 1980s, three scientists (Paul Benioff, Richard Feynman, and David Deutsch) realized that this difficulty could be profitably turned around. Instead of complaining about the problems of simulating quantum mechanics using classical computers, couldn't we build computers out of quantal systems? The richness of quantum mechanics might then allow such "quantum computers" to accomplish more tasks faster than their classical counterparts. For example, in a conventional computer the memory consists of many storage locations that can be set to either "1" or "0", and the processor consists of many switches that can be either "up" or "down". But a quantal system — such as the magnetic needle of a silver atom — can be either "up" ($m_z = +m_B$), or "down" ($m_z = -m_B$), or in an infinite number of other possibilities. Pieces of a quantum computer can interfere or become entangled, options that are not available to the components of classical computers. Can this flexibility be harnessed to make quantal storage locations or switches that work harder than their classical counterparts?

The answer to this question is "yes". For example, in 1997 Lov Grover showed how a quantum computer could outperform a classical computer in searching through an unordered list. Suppose, for instance, that you wanted some information and you knew it was contained in one of ten million possible World Wide Web sites. If a computer could examine one Web site per second, then a classical computer would need on average five million seconds — two months — to find the desired site. A similar quantum computer would find it in forty-two minutes. In 1998 Chuang, Gershenfeld, and Kubinec built a quantum computer that implemented Grover's idea, but the computer could not search through a list of ten million possibilities: it was restricted to lists of four items.

Many issues, both fundamental and technical, must be resolved before the quantum computer becomes more than a laboratory curiosity. Quantum computers may lead society into an information revolution that will make the classical computer revolution look like a ripple. Or the whole idea might just fizzle. But in either case quantum computing illustrates that the quantal domain is fundamentally different from the classical domain, offering up a set of possibilities so various, so beautiful, so new, that they demand a fresh picture of this extraordinary universe, our home.

## B.3   References

A survey of quantum mechanics applications appears in

> Gerard J. Milburn, *Schrödinger's Machines* (W.H. Freeman, New York, 1997).

The role of tunneling in generating sunlight is discussed in

> A.C. Phillips, *The Physics of Stars* (John Wiley, Chichester, UK, 1994) pages 99–100, 110.

Quantum computing is a rapidly changing field and whatever I have said here is likely to be out of date by the time you read this. Nevertheless, I can safely recommend

> Andrew Steane, "Quantum computing", *Reports on Progress in Physics*, **61** (1998) 117–173

as a technical but delightful overview of the whole field as it appeared in 1998, and

> Neil Gershenfeld and Isaac L. Chuang, "Quantum computing with molecules", *Scientific American*, **278** (6) (June 1998) 66–71

as an intriguing general presentation on one experimental realization of the quantum computer.

The specific achievements mentioned above were announced in

Lov K. Grover, "Quantum mechanics helps in searching for a needle in a haystack", *Physical Review Letters*, **79** (1997) 325–328,

I.L. Chuang, N. Gershenfeld, and M. Kubinec, "Experimental implementation of fast quantum searching", *Physical Review Letters*, **80** (1998) 3408–3411.

The last article shows how quantum computers can be constructed using the technique of nuclear magnetic resonance, a technique which evolved out of the Stern–Gerlach experiment.

# Appendix C
## Sources

Page xii, "the belief in an objective reality ...": M. Kakutani, "How technology has changed modern culture", New York *Times*, 28 November 1989.

Page 2, "[the theory of relativity] is a modification ...": Herbert Goldstein, *Classical Mechanics* (Addison-Wesley, Reading, Massachusetts, 1950) page 185.

Page 2, "Nobody feels perfectly comfortable with it.": Murray Gell-Mann, "Is the whole universe composed of superstrings?", in *Modern Physics in America*, edited by W. Fickinger and K. Kiwakski (American Institute of Physics, New York, 1987) page 186.

Page 2, "I can safely say ...": Richard Feynman, *The Character of Physical Law* (MIT Press, Cambridge, Massachusetts, 1965) page 129.

Page 3, The threshold wavelength for stripping one electron from a helium atom: Experiment: S.D. Bergeson *et al.*, "Measurement of the He Ground State Lamb Shift", *Physical Review Letters*, **80** (1998) 3475–3478. Calculation: G.W.F. Drake, "High precision calculations for helium", in *Atomic, Molecular, and Optical Physics Handbook*, edited by G.W.F. Drake (American Institute of Physics, Woodbury, New York, 1996) pages 154–171.

Page 4, "One can popularize the quantum theory ...": N.D. Mermin, *Space and Time in Special Relativity* (Waveland Press, Prospect Heights, Illinois, 1968) page vii.

Page 10, "Above all things we must beware ...": Alfred North White-head, *The Aims of Education* (Macmillan, New York, 1929) page 1.

Page 19, "Sit down before fact as a little child ...": Letter from Huxley to Charles Kingsley, 23 September 1860, *Life and Letters of Thomas Henry Huxley* (D. Appleton and Company, New York, 1901) volume 1, page 235.

Page 41, "spooky": Letter from Einstein to Max Born, 3 March 1947, *The Born–Einstein Letters* (Macmillan, London, 1971) page 158.

Page 75, "unless someone looks …": John R. Gribbin, *In Search of Schrödinger's Cat* (Bantam, New York, 1984) page 171.

Page 111, "The [Heisenberg] uncertainty principle simply tells us …": Hans A. Bethe, "My experience in teaching physics", *American Journal of Physics*, **61** (1993) 972–973.

Page 115, "The single electron *does* interfere with itself …": This passage is from D. Halliday, R. Resnick, and J. Walker, *Fundamentals of Physics*, extended version, fourth edition (John Wiley, New York, 1993) page 1173, but similar statements can be found in almost any introductory physics textbook.

Page 117, "Our imagination is stretched to the utmost …": Richard Feynman, *The Character of Physical Law* (MIT Press, Cambridge, Massachusetts, 1965) pages 127–128.

Page 122, "contemporary reports": See J. Mehra and H. Rechenberg, *The Historical Development of Quantum Theory* (Springer–Verlag, New York, 1982) volume 1, pages 83, 122–123.

Page 122, "I can characterize the whole procedure as an act of desperation …": Letter from Planck to R.W. Wood, 7 October 1931, quoted in J. Mehra and H. Rechenberg, *The Historical Development of Quantum Theory* (Springer–Verlag, New York, 1982) volume 1, page 49.

Page 122, "In a period of most intensive work …": Werner Heisenberg, *Physics and Philosophy* (Harper, New York, 1958) page 31.

Page 123, "At the turn of the year …": J. Mehra and H. Rechenberg, *The Historical Development of Quantum Theory* (Springer–Verlag, New York, 1982) volume 1, page 372.

Page 124, "At Göttingen …": Max Born, "Statistical interpretation of quantum mechanics" (Nobel lecture), *Science*, **122** (1955) 675–679.

Page 124, "This was late spring in Göttingen …": J. Mehra and H. Rechenberg, *The Historical Development of Quantum Theory* (Springer–Verlag, New York, 1982) volume 2, pages 248–249.

Page 125, "One evening I reached the point …": Werner Heisenberg, *Physics and Beyond* (Harper and Row, New York, 1971) page 61.

Page 126, "Göttingen's deluge of formal learning": Letter from Pauli to Ralph Kronig, 9 October 1925, quoted in J. Mehra and H. Rechenberg, *The Historical Development of Quantum Theory* (Springer–Verlag, New York, 1982) volume 3, page 167.

Page 126, "With respect to both of your last letters …": Letter from Heisenberg to Pauli, 12 October 1925, quoted in J. Mehra and H. Rechenberg, *The Historical Development of Quantum Theory* (Springer–Verlag, New York, 1982) volume 3, page 168.

Page 127, "internal periodic phenomenon": Louis de Broglie, *Comptes rendus*, **177** (1923) 507–510, quoted in J. Mehra and H. Rechenberg, *The*

*Historical Development of Quantum Theory* (Springer–Verlag, New York, 1982) volume 1, page 587.

Page 127, "an old girlfriend [from] Vienna": Walter Moore, *Schrödinger: Life and Thought* (Cambridge University Press, Cambridge, UK, 1989) page 194.

Page 127, "discouraged, if not repelled": E. Schrödinger, *Annalen der Physik*, **79** (1926) 734–756, reprinted in E. Schrödinger, *Collected Papers on Wave Mechanics* (Chelsea Publishing, New York, 1927) page 46.

Page 127, "The more I think ...": Letter from Heisenberg to Pauli, 8 June 1926, quoted in Walter Moore, *Schrödinger: Life and Thought* (Cambridge University Press, Cambridge, UK, 1989) page 221.

Page 128, "Quantum mechanics is very impressive. But ...": Letter from Einstein to Max Born, 4 December 1926, quoted in Abraham Pais *'Subtle is the Lord ...': The Science and Life of Albert Einstein* (Oxford University Press, New York, 1982) page 443.

Page 128, "replied by pointing out ...": Niels Bohr, "Discussions with Einstein on epistemological problems in atomic physics", in *Albert Einstein: Philosopher-Scientist*, edited by P.A. Schilpp (The Library of Living Philosophers, Evanston, Illinois, 1949) page 218.

Page 129, "No reasonable definition ...": A. Einstein, B. Podolsky, and N. Rosen, "Can quantum-mechanical description of physical reality be considered complete?", *Physical Review*, **47** (1935) 777–780.

Page 129, "I went to a beer party ...": Richard Feynman, "The development of the space-time view of quantum electrodynamics", *Physics Today*, **19** (8) (August 1966) 31–44.

Page 131, "Visiting Einstein one day ...": John A. Wheeler, "The young Feynman", in *"Most of the Good Stuff": Memories of Richard Feynman*, edited by L.M. Brown and J.S. Rigden (American Institute of Physics, New York, 1993) page 26.

Page 133, "to grasp the situation, to adapt himself to it ...": Charles de Gaulle, *The Edge of the Sword* (Criterion Books, New York, 1960) page 83.

Page 142, "He had discovered ...": John Robison, in the preface to Joseph Black, *Lectures on the Elements of Chemistry* (Mundell and Son, Edinburgh, UK, 1803) volume 1, pages xxvi–xxix.

Page 143, "the greatest problem ...": Michael Horne and Anton Zeilinger, *American Journal of Physics*, **57** (1989) 567.

Page 143, "hallucinations ...": A.W.H. Kolbe, *Journal für Praktische Chemie*, **15** (1877) 473, quoted in J.H. Van't Hoff, *Chemistry in Space* (Clarendon Press, Oxford, UK, 1891) pages 16–18.

Page 143, "a violent irrationality ...": E.T. Jaynes, "Quantum beats", in *Foundations of Radiation Theory and Quantum Electrodynamics*, edited by A.O. Barut (Plenum Press, New York, 1980) page 42.

# Appendix D
## General Questions

Many chapters in this book are followed by problems (see page 10) that pertain specifically to that chapter. This appendix contains questions of a more general character. These questions are designed either to consolidate your understanding or to extend your knowledge. The latter sort of question will require further study, such as through reading books listed in the references. But answering the questions will generally require considerable analytic thought and not just parroting a book from the library.

D.1 *Is God a deceiver?* A central element of René Descartes's philosophy is that we can usually trust our sensual perceptions because God is not a deceiver. The macroscopic world seems to obey the deterministic laws of Newton, yet quantum mechanics maintains that this is just an appearance: the actual laws of physics are probabilistic not deterministic. Does this mean that Descartes was wrong and that God is a deceiver?

D.2 *Is quantum mechanics really strange?* Throughout this account (beginning with its title) I have emphasized that I find quantum mechanics to be strange. My question here: Is quantum mechanics intrinsically weird, or do I find it weird only because of the way I was brought up? For example, in the Middle Ages most people were brought up believing the earth to be flat. The round earth model must have seemed extraordinarily strange to them when it was first broached. (For example, it must have seemed paradoxical that you could travel always due east and yet eventually arrive back at your starting point.) Yet today even children find nothing unnatural about the round earth because they have heard about it from infancy.

Another example comes from chemistry. Joseph Black (1728–1799) discovered carbon dioxide and a number of basic chemical facts.

Soon after Black's death, one of his contemporaries wrote in astonishment that

> He had discovered that a cubic inch of marble consisted of about half its weight of pure lime, and as much air as would fill a vessel holding six wine gallons. ... What could be more singular than to find so subtle a substance as air existing in the form of a hard stone, and its presence accompanied by such a change in the properties of that stone? ... It is surely a dull mind that will not be animated by such a prospect.

Today, few people consider simple chemical reactions to be "singular".

So what's the truth? Is quantum mechanics quite natural, but we were brought up to think otherwise? Or are chemical reactions in fact remarkable, but we were raised in a prosaic era?

D.3 *Layers of explanation.* In section 2.4 (page 9) I argued that the idea of explanation implied explanation in terms of more fundamental ideas, and that the most fundamental ideas could only be described and not explained. It was once thought that these deepest, simplest, most fundamental ideas ought to be "self evident". The fundamental ideas presented in this book have been very far from self evident. Is this a defect in the ideas presented here or a defect in the supposition of self evidence? (From the point of view of biological evolution, does it make sense that our brains should be hardwired to appreciate atomic phenomena?)

D.4 *Learning about quantum mechanics.* Describe your experience of learning about quantum mechanics. What motivated you to read this book? What questions did you have when you started it? Were those motivations satisfied and those questions answered? Did you learn the material by steady accumulation, or were there certain moments ("flashes of insight") when you suddenly came to understand large chunks of material that had been roving unprocessed about your mind? Different people learn in different ways. Which teaching techniques (lecture, conversation, reading, problem solving, film viewing, running computer simulations, etc.) do you think would be most effective for you in learning quantum mechanics? Is this the same answer that you would give for learning about, say, literature? Has this book changed your idea of the concept of "understanding" in science? What is your impression of your current understanding of quantum mechanics? (For example are you confused, disgusted, fascinated, satiated, all of the above?) Which unanswered questions are most important to you? Do you see any way that you can satisfy your continued curiosity?

D.5 *Rephrasing quantum mechanics.* Rewrite a section or a chapter of this book in your own terms. Make it clearer, or more correct, or more interesting than what I wrote. Explain briefly why your version is superior to mine. (Please send the author a copy of your revision and your explanation.)

D.6 *Can all authors be trusted?* In his book *Beyond the Quantum* (Macmillan, New York, 1986) Michael Talbot writes that the Aspect experiment forces the conclusion that "either objective reality does not exist and it is meaningless for us to speak of things or objects as having any reality above and beyond the mind of an observer, *or* faster than light communication with the future and the past is possible". (By the first alternative, he means standard quantum mechanics.) Is either branch of this dichotomy correct, or even internally consistent?

D.7 *What does "fundamental" mean?* Michael Horne and Anton Zeilinger (two of the proposers of the Greenberger–Horne–Zeilinger experiment) write that

> the greatest problem ... is to understand "why quantum mechanics?" Shouldn't a theory as fundamentally important as quantum mechanics follow from something deeper? We suggest that the fundamental elements of quantum mechanics may follow from a careful analysis of what it means to observe, to collect data, and to order them in such a way that physical laws can be constructed.

In section 2.4 of this book (page 9) I took exactly the opposite position, arguing that, by definition, a fundamental theory is one for which such questions cannot be answered. Which position, if either, do you support? Justify your preference.

D.8 *New, bizarre, or both?* In 1877, chemists were just beginning to learn how the arrangement of atoms within molecules could be deduced from chemical information. The distinguished chemist Hermann Kolbe called such attempts "hallucinations ... not many degrees removed from a belief in witches and from spirit-rapping". In 1980, distinguished physicist E.T. Jaynes referred to standard quantum mechanical ideas (such as those presented in this book) as "a violent irrationality ... more the character of medieval necromancy than science". What are your own reactions to quantum mechanics at this stage? Do you believe that Jaynes's reaction is more a rejection of the new and different or a rejection of the irrational? What of your own reaction?

D.9   *Quantum mechanics and Eastern mysticism.* In the 1970s two books appeared concerning the relation between quantum mechanics and mystical aspects of Eastern religion. These were Fritjof Capra's *The Tao of Physics* (Bantam, New York, 1975) written by a physicist, and Gary Zukav's *The Dancing Wu Li Masters* (Bantam, New York, 1979) written by a journalist. Read the two books and compare their treatments of both physics and religion. Can you find any errors in either book? To what extent can the differences in outlook and content of the two books be attributed to the professions of the two authors?

D.10   *Effect of quantum mechanics on culture.* What effect has the discovery of quantum mechanics had on broader human culture, such as philosophy, literature, politics, or popular thought? Are these effects due mostly to quantum mechanics or to misconceptions concerning quantum mechanics?

D.11   *Etymology.* How did the subject of this book come to be called "quantum mechanics"? After all, the word mechanics is usually associated with other activities. (Cartoon below courtesy of Sidney Harris.)

"Actually I started out in quantum mechanics, but somewhere along the way I took a wrong turn."

# Appendix E
## Bibliography

R.P. Feynman, *QED: The Strange Theory of Light and Matter* (Princeton University Press, Princeton, New Jersey, 1985). My favorite book about physics, and the best place to turn if you want to learn more about quantum mechanics after finishing this book. After an inspiring introduction (pages 1–12), Feynman skillfully sets up the framework of quantum mechanics (pages 13–83) and then goes on to give the specific rules — within that framework — for assigning amplitudes for a class of phenomena called "electrodynamics" (pages 83–130). The remainder of the book (pages 130–152) surveys those parts of nature that fall outside of the domain of electrodynamics, and briefly shows how they, too, fit into the quantal framework.

R.P. Feynman, *The Character of Physical Law* (MIT Press, Cambridge, Massachusetts, 1965). Chapter 6, "Probability and uncertainty — the quantum mechanical view of nature", is the best one-hour summary of quantum mechanics that I know. It is the transcript of a lecture that was also recorded on film, and viewing the film is even better than reading the transcript. The video recording is distributed by Education Development Center, Inc.; 55 Chapel Street; Newton, Massachusetts 02158–1060.

P.C.W. Davies and J.R. Brown, *The Ghost in the Atom* (Cambridge University Press, Cambridge, UK, 1986). Interviews with quantum physicists at the popular level.

George Greenstein and Arthur G. Zajonc, *The Quantum Challenge: Modern Research on the Foundations of Quantum Mechanics* (Jones and Bartlett, Sudbury, Massachusetts, 1997). At the level of a junior or senior physics course, this book provides a superb account of experiments, but is sometimes murky in discussing the conceptual consequences of those experiments. For example, the authors say "the [Heisenberg] uncertainty principle prevents us from observing the trajectory of an electron ... in an atom", when they should say "an electron in an atom doesn't have a trajectory, so of course we can't observe it".

145

Jim Baggott, *The Meaning of Quantum Theory* (Oxford University Press, Oxford, UK, 1992). Another clear presentation at the mathematical level of a junior or senior physics course.

Leslie E. Ballentine, editor, *Foundations of Quantum Mechanics Since the Bell Inequalities* (American Association of Physics Teachers, College Park, Maryland, 1988). Reprints of articles, including an excellent annotated bibliography.

Hans Christian von Baeyer, *Taming the Atom: The Emergence of the Visible Microworld* (Random House, New York, 1992). A lyrical popular account. The word "taming" in the title carries the double meaning of "rendering useful" and "rendering familiar and visualizable".

S. Kamefuchi, editor, *Foundations of Quantum Mechanics in the Light of New Technology* (Physical Society of Japan, Tokyo, 1984); Daniel M. Greenberger and Anton Zeilinger, editors, *Fundamental Problems in Quantum Theory: A Conference Held in Honor of Professor John A. Wheeler* (The New York Academy of Sciences (Annals, volume 755), New York, 1995). Proceedings of two conferences devoted to the foundations of quantum mechanics. Much of these two volumes will be incomprehensible to the non-physicist, but they will show you that seasoned professionals as well as neophyte amateurs are fascinated and confused by the issues raised in this book.

## Popularizations

Below are six conventional popularizations of quantum mechanics. I feel that each one of them is deficient either through oversimplification or through an emphasis on people rather than on nature. However, the authors of these books probably feel the same way about mine.

J.M. Jauch, *Are Quanta Real?* (Indiana University Press, Bloomington, Indiana, 1973). In many ways the best of the popularizations, but written before the full significance of the Einstein–Podolsky–Rosen paradox was understood.

Heinz Pagels, *The Cosmic Code: Quantum Physics as the Language of Nature* (Simon and Schuster, New York, 1982). An attempt to cover quantum mechanics, special and general relativity, statistical mechanics, elementary particle physics, and the history of each of these fields, all in one volume.

John R. Gribbin, *In Search of Schrödinger's Cat: Quantum Physics and Reality* (Bantam, New York, 1984). Describes both quantum mechanics and its history. Contains a few errors (pages 8, 167, 171, 229, 261, 265, and two on page 176). The author occasionally uses bizarre and misleading terminology, such as "the electron is not real" when he means "the electron does not have a definite position".

John R. Gribbin, *Schrödinger's Kittens and the Search for Reality: Solving the Quantum Mysteries* (Little, Brown and Company, Boston, 1995). Some history, some experimental tests, some alternative interpretations. Occasionally oversimplified to the point of error, as when, on pages 92–98, "probability" is used to mean "amplitude".

J.C. Polkinghorne, *The Quantum World* (Princeton University Press, Princeton, New Jersey, 1984). Written by a physicist turned priest. Nice description of the mathematical tools used by physicists to squeeze results out of the quantum theory. To my mind, this is more a book about how humans study nature than a book about nature, but if you want to find out what eigenvalues are and why you should care about them, then this is the book for you.

Alastair Rae, *Quantum Physics: Illusion or Reality?* (Cambridge University Press, Cambridge, UK, 1986). Contains six errors on the first three pages, but then improves.

## *Effect of quantum mechanics on culture*

Caution: Some of these authors get the physics wrong. Some of them can't even distinguish between quantum mechanics and the theory of relativity!

Tom Stoppard, *Hapgood* (Faber and Faber, London, 1988). A sophisticated spy play involving quantum mechanics.

John Barth, *On With the Story* (Little, Brown and Company, Boston, 1996). In "Love Explained", one of the short stories in this collection, a character maintains that "More than Freudian psychology, more than Marxist ideology, quantum mechanics has been the Great Attractor of the second half of this dying century — even though, speaking generally, almost none of us knows beans about it."

"Wavefunctions for String Trio (four vignettes about the new [quantum] physics)" by John Tartaglia. Recording by Ensemble Capriccio published in 1999. The vignettes include "Bell's theorem" and "Schrödinger's cat".

Jane Hamilton, "When I began to understand quantum mechanics", *Harper's*, **279** (August 1989) 41–49. A short story involving quantum physics and a beauty pageant.

June Jordan, "Poem on the quantum mechanics of breakfast with Haruko", *The Nation*, **257** (5 July 1993) 40. A love poem.

Eric Kraft, *Where Do You Stop?* (Crown, New York, 1992). Cataloged by the Library of Congress under "Quantum theory — Fiction".

Robert Anton Wilson, *Schrödinger's Cat* (Simon and Schuster, New York, 1979). A novel.

Susan Strehle, *Fiction in the Quantum Universe* (University of North Carolina Press, Chapel Hill, North Carolina, 1992).

Robert Nadeau, *Readings for the New Book on Nature: Physics and Metaphysics in the Modern Novel* (University of Massachusetts Press, Amherst, Massachusetts, 1981).

Aage Peterson, *Quantum Physics and the Philosophical Tradition* (MIT Press, Cambridge, Massachusetts, 1968).

Lawrence Sklar, *Philosophy of Physics* (Westview Press, Boulder, Colorado, 1992).

Jonathan Powers, *Philosophy and the New Physics* (Meuthen and Company, London, 1982).

James T. Cushing and Ernan McMullin, editors, *Philosophical Consequences of Quantum Mechanics: Reflections on Bell's Theorem* (University of Notre Dame Press, Notre Dame, Indiana, 1989).

# Appendix F
## Skeleton Answers for Selected Problems

Be sure to read page 10 about the philosophy behind active learning and problem solving before using these skeleton answers.

**2.1.** Large force directed downward, small force directed downward.

**2.2.** $A > B = D > C$.

**2.3.** 2800 miles.

**2.4.** $-0.38$ inches.

**2.5.** Infinite number, all perpendicular to the arrow.

**3.1.** All of the atoms would leave at one deflection corresponding to a large positive projection.

**4.1.** (a): 3 inches, (b): $-3$ inches, (c), (d), and (e): 0 inches, (f): $3/\sqrt{2} = 2.121$ inches.

**4.2.** They would all leave the $-$ exit.

**4.3.** Because of the qualifier "in general", the claim is consistent with situations in which the probability of one outcome is 1 and the probability of all the other outcomes is 0.

**4.4.** (2).

**4.5.** All $-$. Half $+$ and half $-$.

**4.6.** Not at all.

**4.7.** No.

**4.8.** $3/4$.

**4.9.** $1/2$.

**4.10.** $3/4$, $1/4$.

**4.13.** In both cases, "It just *is* correct. I can tell you about experiments which show *that* it is correct, but I can't say *why* it is correct."

**5.1.** 1/72.

**5.2.** 1/10.

**5.3.** 5/6.

**5.4.** (a), (b), and (c): $1/2^{10}$, (d): $10/2^{10}$.

**5.5.** (a), (b), and (c): 1/2.

**5.6.** (c): 5/8, 6/8.

**5.7.** Hint: "Thirty days hath September ...".

**5.9.** (a): $1/(25 \times 10^{12})$, (b): $1/(5 \times 10^{6})$, (c): $51/(5 \times 10^{6})$, (d): $(52 \times 51/2)/(5 \times 10^{6})$.

**6.3.** 1/4.

**6.4.** 7/9.

**8.1.** Length 12.07 inches, direction 1:30 or "northeast".

**8.2.** Either "all of them" or "none of them" are acceptable answers.

**9.1.** These phenomena happen even when only one atom is present in the apparatus.

**9.2.** (a): 1/4, (b): 1/4, (c): 1.

**9.3.** 50%, 50%, 0%, 100%, 50%, 0%, 50%, 0%, 12.5%.

**9.6.** (1) Measurement means someone looks. (2) An electron is a marble with a definite position, that goes through one hole or the other but neither you nor nature knows which.

**10.2.** If an atom's position were always definite, then quantal interference (experiment 9.3) would be much worse than a puzzle, it would be a logical contradiction. We are able to regain logical consistency only by abandoning the mental picture of an atom as a small, hard marble.

**11.2.** $1/\sqrt{2}$, $1/\sqrt{2}$, 0.

**11.4.** 1/4.

**14.1.** Yes, yes, no, no, no.

**15.1.** "Between release and detection, the electron is not at *any* point, because it doesn't have a position. Instead, it has amplitude to be at each of many points."

# Index

Aharonov–Bohm effect, 94–95, 97
AIDS, 20
aim of this book, xi–xiii, 11, 98, 105, 133
Alice, 99
amplitude, xi, 76–93
analyzer
 front–back, 86
 Stern–Gerlach, 22
 tilting, 33, 35
 $y$, 86
analyzer loop, 64
arrows
 addition of, 61
 for amplitude, 77
 for magnetic needle, 5
art, 55
Aspect experiment, 38, 41, 46, 49, 129
auto mechanics, xii, 144
axes, figure for, 22

Bell's theorem, 41, 46, 129
 test of, 41
Bell, John S. [1928–1990], 46, 129
Bethe, Hans [1906–     ], 111
bit, 99
Bob, 99
Bohm, David [1917–1992], 45, 46, 64, 78
Bohr magneton, 15
Bohr, Niels [1885–1962], 1, 18, 123, 128
Bohr–Einstein debate, 128

Born, Max [1882–1970], 124, 128
brink of implausibility, 41

chess, 4, 30, 85
classical limit of quantum mechanics, 16, 69, 83, 103–106, 111, 141
classical mechanics, 1
clockwork mechanism, 27, 41, 63, 68, 70, 117
codes, 98
common sense, xii, 3, 83
compass needle, 5, 16
complex number, 61, 115
compound probability, 32, 66
computer mail, xiv, 99
computer programs, 19, 46, 118
computers, quantum, 20, 134–137
conspiracy theory, 96
contact the author, xiv
conundrum of projections, 21, 26, 28, 85
corkscrew, 94
correspondence principle, 16
cryptography, quantum, 98–102

Davisson, Clinton [1881–1958], 107
de Broglie relation, 105
de Broglie, Louis [1892–1987], 1, 127, 128
definite value
 use of, 24–26, 38, 43, 51, 57, 66, 74, 87, 88, 103, 105, 107, 112, 116, 133
 warning about, 26, 87, 95
delayed choice experiments, 96–97, 108